HYDRATION
AND
BIOLOGY

LEWIS H. FLINT

Foreword by
STUART HALE SHAKMAN

AUTHOR'S NOTE

Chemical science has enjoyed a world of challenge in appraising the nature and inter-relationships of the elements, but for an emergent biological science it is especially important to consider the relationships of the elements and their compounds to an aqueous solvent. Hydrational potentiality is the ionic attribute which organic nature, from its earliest amoeboid beginnings to its present production of astounding types of intellectuals, has used extensively in its embodiment of the precious fire called life. To biology as the life science hydration thus is a matter of most intimate and timely concerns.

Lewis Herrick Flint

circa 1917-1918 circa 1960

CONTENTS

LIST OF FIGURES

Foreword
by Stuart Hale Shakman

"It's mind-boggling." said Dr. Ponnomperuma.

The late Dr. Cyril Ponnomperuma of the U. of Maryland was an authority on the origin of life. When he was reached via telephone in July/Aug., 1983, he spoke of how he and other scientists had been investigating the origins of life in the laboratory over the past 25 years, trying to recreate the conditions that had existed billions of years ago. Their assumption had been that these conditions no longer existed. But Dr. Ponnomperuma nearly bubbled over in excitement as he described what he believes is occurring at deep-sea vents, i.e.:

" ... those conditions that existed then exist now."

- - - - - - -

The incredible set of circumstances that had enabled Lewis Flint's discovery of the principles of hydration – what must be considered "the secret code of the universe" – were substantially documented in his 1964 book, *Behavior Patterns of Hydration*, and exposed in footnotes of his original 1932 *Journal of the Washington Academy Sciences* articles on the subject. As further summarized in Flint's subsequent 1973 book, *Dissenting Ape*:

"Had it not been for the stock market crash of 1929 I would not have been loaned out [by one US government office to another] and assigned to research involving electrical measurements. Had I not been curious about the significance of what I was doing I would not have become interested in the behavior of the K, Na, and Li ions in an electrical field. Had I not been abstracting French and German articles for Biological Abstracts I would not have been reading an article in a German periodical in the library. Had I not been studying the behavior of the K, Na, and Li ions at the time, I would not have paid any attention to the article entitled "Electro-Affinity—a New Force in Chemistry [Abegg &Bodlander, 1899]," which dealt with the hydration of the ions K, Na, and Li. Finally, had I been a chemist I would in all probability never have had the nerve to project a description of hydrational potentiality so vitally at variance with concepts long held and revered as factual in chemical science."

Apparently, this German article had sent Flint back to his 1915 *Principles of General Physiology* textbook by W.M. Bayliss (presumably carried over from his Harvard graduate school studies) to review data concerning the K+, Na+, and Li+ ions. Here were listed conductivity measurements by Nernst, and on a consecutive page, proposed hydration numbers by Bousfield (which seemingly-essential data would be missing from the next edition of Bayliss's textbook!).

- - - - - - -

Unlikely as might be considered the essential sequence of events underlying and enabling Flint's great discovery, my pathway to the very consideration of Flint's great work seems equally if not even moreso serendipitous. While living in the seaside community of Venice in southern California, I had been developing strap-on roller skates with polyethylene wheels, and intending to write an article on the history of outdoor roller-skating. A test ride of my latest prototype through the center of "town" was obstructed by a small convoy of big trucks with ground-shaking vibrators extended down from their bottoms, and from which long strings of wires with regularly-spaced devices (so-called "geo-pods") were deployed along the ground. The trucks set off vibrations that felt like little earthquakes. Inquiring as to their purpose, a worker responded, "Listening for oil".

Coincidentally, long afterward I encountered a 1980 *Scientific American* article which discussed the possibility of the existence of unlimited resources of oil and gas, particularly natural gas. The article summarized the mounting evidence that oil and gas deposits around the globe may be natural phenomena of our evolving earth, and not the product of decaying fossils: "Diverse evidence leads us to believe that enormous amounts of natural gas lie deep in the earth, and that if they can be tapped, there would be a source of hydrocarbon fuel that could last for thousands of years." [*Scientific American*, June 1980, "Deep Gas Hypothesis" by Thomas Gold and Steven Soter.]

The confluence of the local oil/gas exploration activities and global implications of the Gold/Soter article diverted my attention through 1981, to the extent time allowed, to considerations of the quantities and origin of "fossil fuels" world-wide. Particularly eye-opening was a local newspaper article, *Evening Outlook*, Jan 29, 1982, page a-4, entitled "Undersea 'petroleum factories' discovered". The article discussed an expedition to superheated hot springs at at the bottom of the mile-deep Guaymas Basin, about about 35 miles off the Mexican coastal city of Guaymas in the Gulf of California and 600 miles southeast of San Diego. At these hot springs on the ocean floor, the expedition had discovered what was described as apparently "juvenile" deposits of petroleum

"The expedition also reported finding the largest bacteria ever discovered – about the size of a grain of sand – in huge mattings up to three deep on rocks and the sea floor."

In an attempt to learn more, I visited Scripps Institution at U.C. San Diego in March 1982. There I was able to meet and discuss, with Dr. Harmon Craig, the methane gas that is being emitted from hydrothermal vents on the ocean floor. Dr. Craig told of his conclusions concerning the non-biological ("abiogenic") origin of this methane.

I asked if any of the methane coming from the vents might be reaching the atmosphere.

"No," said Dr. Craig. "It's all being consumed by bugs. ... Billions of bugs outside the vents. Swarms of bugs."

"Bugs?" I asked. "Yes, bugs." he replied. "Bacteria. It must be bacteria." he offered in clarification.

In an other-worldly tone of voice, Dr. Craig described the mushroom-shaped-cloud appearance of a hydrothermal plume as it virtually explodes upward from a vent and spreads laterally as it reaches the density of the surrounding ocean.

These volcanic hot gases, pushing into to the already-totally-compressed waters on the the ocean floor, pose a classic situation of two immovable objects merging, i.e., 10,000 degree gases that cannot be further compressed, being folded into the solid floor of 4 degree water that likewise cannot be further compressed. The inevitable result is that both substances are somehow forcibly dissociated into their respective component parts, ions if you will, and in-so-doing are inverted into a type of bubbles containing the now-dissociated and energetic, vibrating, ions. These "bubbles" seemingly comprise the simplest of one-celled organisms, or bacteria, falling into the ocean floor in mats. Yes, life is evolving on the ocean floor before our very eyes, today as it did billions of years ago.

Following up on April 23, 1982, I was able to reach Sidney Fox at the University of Miami. I asked, "Is the origin of petroleum and origin of life related?

"I think so" he said.

In a letter, April 27, 1982, Dr. Fox suggested that at the point of origin (possibly pre-existing) proto-amino acids could go toward life or be cooked out into petroleum", this aside from question of the source of these proto-amino acids (i.e., whether these were pre-existing or the result of spontaneous generation).

Dr. Fox's letter provided copies of relevant background information: "Enclosed in a 1957 paper is a suggestion of hydrothermal origins of precursors to life. This is now considerably strengthened by findings in the Galapagos Rift, Spirit Lake, and the East Pacific Rise".

"The complex interfaces are many. The main picture looks like the prebiotic Earth had to have layers of amino acids. In hydrothermal vents, these could have accumulated. At a reasonable state of cooking, the polymers of amino acids would have formed, been ejected into marine environment (there are some special steps involving formation of basic and acid polymers), and the aggregation of polymer into protocells. This view is documentably 'generally accepted' (Follman, Naturwissenschaften 69, 75-81; 1982). The portion of amino acids that did not go this chemical route, toward life in some cases, would dry out and be decomposed to petroleum."

> [Sidney W. Fox, personal correspondence 27 April 1982; *Journal of Chemical Education*, Vol 34, 72, Oct. 1957; *Reprint of The Fifth World Petroleum Congress*, June 4 1959, Fordham, NY, Sidney W. Fox and G.G. Maier, p. 9, "The production of petroleum as a by-product of in generation of life deserves consideration."]

Thus, in some cases, where these bacteria are caught in the upwelling of the emerging hot gases, they are being cooked out by the extreme heat and become another substance – what we call petroleum. And we can begin to consider the possibility that huge petroleum deposits found in various locations around the world are the result of similar or identical such processes under some ancient seas, covered over by various geological processes and the sands of time.

On or about April 23-25, 1982, this phenomenon was discussed with Sol Silverman, the expert at Chevron, via telephone. It was noted that *Petroleum Geology* by Tissot Welte, 1978, p. 131, cited John Hunt of Woods Hole, Mass., as the source of calculations based on the Arrhenius equation for petroleum generation. According to Hunt, oil formed in 25 million years at 110 degrees C would take 50 million years at 90 degrees C. Conversely this same oil would form in 24,414 years at 210 degrees C, 24 years at 310 degrees C, 1 ½ years at 350 degrees C, and 17 days at 400 degrees C. As I recall, when presenting these calculations to Sol Silver, he said "I doubt it would take that long."

In other words, at the types of extreme temperatures and pressures encountered at the interface of undersea volcanoes and the ocean floor, given the proper mix of ingredients, the generation of generation of oil could be virtually instantaneous.

Regarding these "bubbles", or bacteria if you will, it is the phenomenon of their coming into existence that motivated this writer to seek an understanding of the generic relation between *biology* and *hydration*, how life's origin itself seems to the result of the process of hydration – the process whereby these incredibly-hot gases spewing forth from under-sea volcanoes are somehow folded into the immovable and non-compressible waters on the ocean floor.

It was through Harmon Craig of Scripps, and his wife Valerie, that I had learned of the upcoming May 3-4, 1982, Deep Source Gas Workshop in Morgantown W. Va. This was the impetus for cutting short my (four years') tenure in Venice/Los Angeles. Gathering up whatever financial resources could be raised, arranging for a east-bound drive-away car, packing up what belongings that would fit in the car with my string bass strapped to the roof, I was off into a rare L.A. rain storm – destination Washington, D.C. via Morgantown W. Va. I dropped the bass in Chicago for safe-keeping, hopped a bus from the car-drop-off point in Ohio to Morgantown,W.Va., and checked into the conference.

It seems that almost everyone there (except for me) was either from the oil/gas industry, an academic institution, or the government. It was a pretty technical group, but the essential message concerned the existence of "deep source gas" – how deep was it being found and what was its suspected source – i.e., was it from decaying dinosaurs etc. (so-called "fossil fuels"), or was it "abiogenic", left over from the earth's formation. It is safe to say that the consensus was that a substantial portion at least was indeed "abiogenic", and indicative of large, virtually inexhaustible, deposits, but actually this was not the only big take-away from the conference.

The surprise package was the existence of a substance called "methane hydrates", in huge quantities, underlying the permafrost and offshore regions of the world's oceans. The first acknowledged existence of this substance came from a problem experienced with above-ground transmission lines of natural gas. These had been getting clogged when the natural gas being transmitted contained water, even above the freezing point of water. The combination of water and gas, under pressure, formed this substance, a structure, that formed a solid cage-like ("clathrate") structure generally containing 23 water units and 6 gas units (there was some variability, but this was the general idea). Of course the simple solution was to keep the gas lines free from water.

Independent of the gas line problem, gas and petroleum exploration programs had more recently identified large deposits of methane hydrates beneath under the permafrost and offshore. The estimates of the quantities of gas that might be contained in these deposits are astronomical – the high estimates, attributed to a Russian, a Mr. Dobrynin, as presented by Dick McIver of Houston TX (and confirmed in telecom a/o May 30, 1983), was 2.7×10^8 TCF. If we calculate for comparison the quantities of air in the atmosphere, assuming a approximate 10-mile thick blanket air surrounding the earth, there would be four times as much methane in hydrates as there as air in the atmosphere.

[Dobrynin, VM, etal., "Gas Hydrates – One of the Possible Energy Sources", Preprints from the IIASA Conference on Conventional and Unconventional World Natural Gas Resouces, 4th Conference, Montreal, Canada (1979)]

Moreover, these huge deposits of methane hydrates may (indeed probably) serve as caps for huge quantities of gas trapped beneath them. Indeed, rather than looking at natural gas as an inexhaustible natural resource for exploitation, it was looking more and more like gas itself was a geological "fact of life" of the earth itself.   "Solid, ice-like mixtures of natural gas and water have been found immobilized in rocks beneath the permafrost in Arctic basins and in muds under deep water along the continental margins of the Americas. While these unusual compounds have been known and studied by chemists for over a century and a half and by gas transmission engineers for the past 50 years, the discovery of natural occurrences came as a surprise to earth scientists. ...

"Hydrates are solid ... and may even form seals and traps that will trap free gas where the structure and stratigraphy of the rocks themselves would not."

[From Rodney Molone, I learned of a recent workshop on this subject and was able to obtain a copy of DOE/METC 82-49 Methane Hydrates Workship Technical Proceedings March 29-30, 1982, "The Occurrence and Magnitude of Methane-Hydrate Accumulations by R.E. Zielinski and R.D. McIver]

- - - - - - -

At the end of the conference I was able to hitch a ride to Washington, DC, and my "home away from home" for the next year and a half, the Library of Congress. Once there I found a title in the card catalog that fit my search criteria, the terms *"hydration"* and *"biology"*, in Flint's 1968 book *Hydration and Biology*. This led to Flint's earlier book, *Behavior Patterns of Hydration*, which authoritatively established that Flint had indeed uncovered the "holy grail". In his words: .

"At the time of the discovery in 1932 I had attained the age of 39, an age held and maintained by no less an authority than Jack Benny to represent the very peak of perfection in a human male." L. Flint, *Behavior Patterns of Hydration*, 1964.

This current volume, *Hydration and Biology*, explores further physiological implications of Flint's discovery, providing particularly-detailed long-standing experiments involving osmosis. Here we begin to probe the magic that enables the miracle of life to be expressed in its various forms. In this Foreword are pointed out some associated and notably-broad other implications:

(A) Ponce de Leon's Fabled Fountain of Youth highlights the potential power of Flint's methodology relative to life's processes.

(B) "Beam me up Scotty" suggests the potential of the realization of a four-dimensional algebraic calculable universe.

(C) Prout's Hypothesis Reborn points out how Flint's work establishes the validity of Prout's hypothesis that all elements are composites of a basic one.

(D) Relations Between Periodicities of Mendeleev and Flint – Mathematical & Structural:

    (1) Mathematical – Law of Gravitation on the Ionic Level illustrates how Mendeleev's periodicity is apparently based on ionic weight, or gravitation.

    (2) Structural – A Combined Periodic Table of Elements, incorporating Mendeleev and Flint's taxonomies, reveals inherent underlying symmetries.

- - - - - - -

(A) Ponce de Leon's Fabled Fountain of Youth

Flint's experiments involving osmotic measurements provided not only a wealth of validating data, but thereby provided the framework for further extension of related studies into biology, as emphasized in this volume. As independently emphasized by Gilbert Ling, it is only through an understanding of osmotic force can we come to understand the mechanism behind the mechanical force of movement that makes the plant and animal world possible.

Some indication of the significance of this line of inquiry is evident in some excerpts from Gilbert N. Ling's *In Search of the Chemical Basis of Life*, in which he states the "...conclusion that the widely accepted membrane pump theory of the living cell is incorrect." [p. xxvi] … Ling discusses an approach "wherein the contractile force depends on an osmotic gradient... How precisely this water activity reduction affects contraction is not known and may prove to be a fruitful area for future research." [p. 572]

Gilbert Ling: "How is [the] potential energy trapped in the resting muscle made to perform mechanical work? ... it is postulated that localized changes in the osmotic activity of water (concomitant with the release of K+ ions in one area and depolarization of water in another) provide the major force for muscle contraction." [p. 582-584]

As noted throughout the work of Lewis Flint, this is not a replay, but a new tool for truly beginning to understand the source of energy in metabolism; call it the "mathematics of metabolism". At the same time

Flint emphasized that the suggested lines of inquiry that he explored are not in and of themselves definitive. Rather, they are merely a starting point, or starting points, for further exploration.

Given the recent and accelerating strides in our abilities to assess great quantities of data through tremendous advances in computing capabilities, the possibility of taking the simple algebraic foundation uncovered by Flint to near-unimaginable places is truly mind-boggling.

If comprehensible and algebraically calculable osmotic force is the essential force that powers life's processes, then we are faced with the likelihood, not merely the possibility, of attaining the ability to gain a far greater degree of control over these processes than previously imagined, other than in science fiction stories. If life's processes are merely some combination of algebraically calculable electro-chemical changes, the prospect of even reversing such processes, e.g. processes of aging and disease, through a reversal of these electro-chemical changes seems not so impossible, or even nearly within grasp.

(B) "Beam me up Scotty"

A memorable line from a contemporary futuristic show, "Beam me up Scotty" may find its algebraic foundation in a merger of Flint's and Moseley's respective works relative to the atomic number. At a minimum, Flint's juxtaposition of atomic number versus electrical conductivity may be viewed as an independent confirmation of Moseley's landmark work.

Moseley established that the atomic number varies with the square-root of the frequency of vibration, or inversely with the square-root of wavelength. Flint showed how solute ionic number (+ valence) varies with the inverse square of conductance, which seems to be doing somewhat of the same thing from a different direction, i.e., validating the primacy of the atomic number. At the same time, the overall scheme establishes some precise and definite correlation between conductance and frequency, i.e., wavelength varies with conductance to the fourth power, i.e. a four-dimensional correlation wherein the atomic number is sandwiched between inverse squares.. This may be readily illustrated in tabular form:

Table 1   Figure 1

L	W	Valence C	Calc.Cond. EC-calc 545.3/sqrt: ((81/sqrt:WL) +7.5+C)	EC-obs	Error
La	2.676	3	70.388783	69.7	-0.00979
Ce	2.567	3	69.786526	69.8	0.000193
Pr	2.471	3	69.237182	69.5	0.003796
Nd	2.382	3	68.710911	69.4	0.010029
Sm	2.208	3	67.630331	68.5	0.012859
Eu	2.13	3	67.121645	67.8	0.010106
Gd	2.057	3	66.630678	67.3	0.010045
Dy	1.914	3	65.623457	65.6	-0.00036
Er	1.79	3	64.695961	65.9	0.018611

Figure 1 — EC-calc v. EC-obs.

(C) Prout's Hypothesis Reborn

In 1815 William Prout advanced the hypothesis that the atomic weights of the various elements are exact multiples of the weight of the smallest element, hydrogen. Notwithstanding the painstaking efforts of armies of chemists through the 19th and 20th Centuries, up to the present time, to establish definitive atomic weight values based on relative calculations involving combinations of elements in compounds, conventional chemistry still, even as recently as the year 2014, continues to juggle alternate sets of irregular values for atomic weights.

The work of Flint in particular, in conjunction with a number of other independent perspectives, provides seemingly inconvertible support for the solidity of Prout's position; Flint's hypothesis of a weight of 4 for H2 gas on a scale of 32 for O2 gas* is supported by

(1) similarities with He gas:

  (a) He has about 98% the lifting power of H2** and

  (b) an average of less than 4% separates observed diffusion coefficients for He and H2 in 8 instances affording direct comparison listed in references***;

(2) peculiarities in the behavior of H2 gas as noted by

  (a) Arrhenius, who referred to hydrogen's cathode ray absorption-to-density ratio (5610) as a "notable exception" from the mean [2794=mean for 8 sol- ids (collodium, paper, glass, mica, aluminum, brass, silver, gold) and 3 gases at 760 mm. Hg. (H2, air, SO2) as derived by Lenard];

however, when H2 is assigned a weight of 4 (v.s. 2), its absorption-to-density ratio (2805) correlates closely with this mean****, and

(b) Graham, who commented that the "want of mechanical equivalency in hydrogen mixtures is exceedingly remarkable, being a marked departure from the usual uniformity of gaseous properties"*****; and
(3) previously reported studies involving approximately 200 diffusion coefficients******

*FLINT, L.H., Behavior Patterns of Hydration (1964), Ch. 11
**CRC Handbook of Chemistry and Physics, 1985-6, B-20.
***CHAPMAN, S., etal, Mathematical Theory of Non-Uniform Gases (1970), 263, 267; and HIRSCHFELDER, J.O., etal., Molecular Theory of Gases and Liquids (1954), 579, 601.
****ARRHENIUS, S., Theories of Chemistry (1907), 91.
*****GRAHAM, THOMAS, Elements of Chemistry (1850), 81.
******SHAKMAN, SH, Abstracts AAAS, 1986, p. 119 (No. 212).
*******SHAKMAN, SH, Abstracts AAAS, 1987 (No. 110), "Observations of Behavior of H2 Gas"

(As Newton affirmed,"More is in vain when less will serve."; see also *Principles of Hydration* – Uni-Science abstracts/ associated calculations.)

(D) Relations Between Periodicities of Mendeleev and Flint – Mathematical & Structural

(1) Mathematical – Law of Gravitation on the Ionic Level

In Figure 2, the maximally-hydrated atomic numbers are arranged sequentially, from the smallest (23+0) to the next (22+9; 9 equals the atomic-number-plus-valence-equivalent of the hypothesized negatively-ionized H20- ion), and so on. Once so-arranged (in 5 columns), a box is drawn around the ions within each Mendeleev grouping. Therein is seen, as for example, *in each of three Mendeleev Groups (I, II, III, repectively), three ions (with atomic numbers 3, 29, and 55; 4, 30, 56; and 5, 31, 57; resp.) have nearly-identical/consecutive Z'h values* (maximally hydrated atomic number equivalents), i.e., nearly-identical "weight" values. The unlikihood of such groupings being random suggests a relation between the respective and similar chemical behavior in aqueous solution, i.e. that on which Mendeleev's periodicity is based, and their respective maximally hydrated weights. In other words, this is suggestive of the *action of gravity on the ionic level* as regards the cited ions in respective groupings.

Figure 2: Hydrated Atomic-Number per Flint (Zh) versus --- Mendeleev Periods

Zh(Z)	Zh(Z)	Zh(Z)	Zh(Z)	Zh(Z)	
		148(85)	212(77)	276(69)	
	85(67)	149(59)	213(51)		------------ (V)
	86(41)	150(33)	214(25)		
23(23)	87(15)	151(7)			
	92(92)	156(84)	220(76)		
	93(66)	157(58)	221(50)		------------ (IV)
	94(40)	158(32)	222(24)		
31(22)	95(14)	159(6)			
	100(91)	164(83)	228(75)		
	101(65)	165(57)	229(49)		----------- (III)
	102(39)	166(31)	230(23)		
39(21)	103(13)	167(5)			
	108(90)	172(82)	236(74)		
	109(64)	173(56)	237(48)		------------ (II)
46(46)	110(38)	174(30)			
47(20)	111(12)	175(4)			
	116(89)	180(81)	244(73)		
	117(63)	181(55)	245(47)		----------- (I)
54(45)	118(37)	182(29)			
55(19)	119(11)	183(3)			
	124(88)	188(80)	252(72)		
	125(62)	189(54)	253(46)		
62(44)	126(36)	190(28)			
63(18)	127(10)	191 (2)			---------------------- (0)
	132(87)	196(79)	260(71)		
69(69)	133(61)	197(53)			---------------------- (VII)
70(43)	134(35)	198(27)			
71(17)	135(9)	199 (1)			
	140(86)	204(78)	268(70)		
77(68)	141(60)	205(52)			---------------------- (VI)
78(42)	142(34)	206(26)			
79(16)	143(8)	207(0)			

Z = Atomic number; Zh = Z+9Hmax; Hmax = 23n-Z, and n=1 when Z= 0-23, n=2 when Z=23-46, n=3 when 46-69, and n=4 when z=69-92.

[Note: Above calculations assume neutral values for unhyrated atoms.]

(2) Structural – Combined Periodic Table of Elements, Mendeleev/ Flint

This combined periodic table of elements* preserves essential information of both taxonomies and may facilitate approaching scientific information from the (alternate or complementary) perspective of either.

The utility of Flint's taxonomy serves to emphasize the significance of atomic number as a "more fundamentally important property of an element than its atomic weight" [J.C. Speakman, 1947]. Here, atomic number (rather than atomic weight) is utilized as a measure of relative weight. This combined table of elements

(a) may further be seen to relate to origins of Mendeleev's periodicity in its symmetrical embodiment of the non-metals and the otherwise somewhat anonymous lanthanides & actinides;

(b) may be viewed as geometrically harmonious with the work of R.B.Fuller; and

(c) constitutes a proposed extension/merger of conventionally-appraised periodicities into an encompassing algebraically-structured system which underlies the fabric of the ponderable universe. The existence of such a system was anticipated by both I.Newton** & A.Einstein***

*SHAKMAN,S.H., Proceedings,Pac.Div.AAAS,Vol.6,Part 1,p.39.
**NEWTON,Principia(1687). ***EINSTEIN,Relativity(1956)

Figure 3

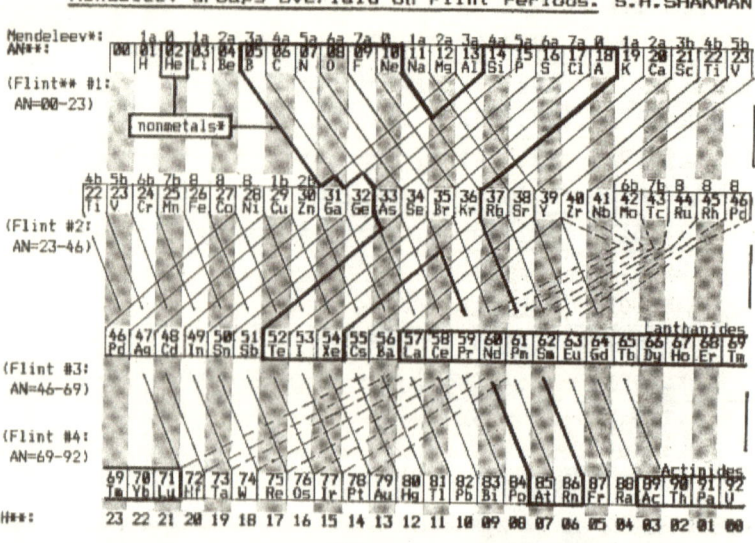

Mendeleev Groups overlaid on Flint Periods. S.H.SHAKMAN

*MORTIMER,CE,Chemistry(1975). **FLINT,LH,Behavior Patterns of Hydration(1964), 21:
H=23n-(AN+-C) [H=(Max.)Hydration No.; n=Period (#1-4); AN=Atomic No.; C=valence].

PREFACE

While as a science biology is still having its birth pains, botanists and zoologists who have attained the age of contemplation may find some measure of consolation in having preceded the contemporary projected administrative orgies of RNA and DNA as writhing helical molecules. Yet seeming complexities always have represented products if not triumphs of natural evolution. At times these complexities have been attributed to or associated with imagined supernatural forces having no validity in science. With the departure of vitalism, however, science became fully aware of its responsibilities and their challenge. In science, nature became no longer a forbidden area.

When physics and chemistry meet within an aqueous solvent, and solutes enter, the stage is set for drama in both the inorganic and the organic worlds. It is the drama of the organic world which is biology, and on its stage there unfolds with ever-increasing complexity and enchantment the magic of life. An indispensable part of this magic basically is involved with of an attribute of light ions which has been described as an hydrational potentiality- an ability to remove from moisture and to take on and hold definite characteristic numbers of H_2O^- units. Eventually biological science, even though it be for no other reason than a desire for a change, is certain to include considerations of the behavior patterns of relatively simple atoms and molecules, ionic and hydrating within an aqueous medium. A preceding volume entitled "Behavior Patterns of Hydration" was devoted to such considerations. In the present volume the same ionic hydrational attribute is utilized as a prospective pathway to an extension of new and old salients on the frontiers of biological knowledge.

While the text of the first volume was in press three supplementary papers were published in Advancing Frontiers of Plant

Sciences, Volumes 6, 7 and 8. Other papers had been prepared for the same journal when it seemed appropriate to group them in the present volume, since they all dealt with subjects of interest in biology. The three published papers have been included herewith as Chapters 8, 9 and 10.

CHAPTER 1. INTRODUCTION

In a general way biology, the life science, might be appraised as a relatively new science which had emerged in substantial measure from botanical science, concerned with plants, and zoological science, concerned with animals. Although plants long had been recognized as the only structures within which chlorophyll characteristically captured radiant energy and elaborated food, upon which all animals as well as all plants were directly or indirectly dependent, it also had been recognized that plants and animals had many common aspects and behavior patterns, especially in the area of protoplasmic metabolism. In retrospect it had been apparent that throughout the long geological period embracing the evolution of the organic world there had been a contemporary interdependence: at the outset interdependence among the simple or less-differentiated plants and animals, and thereafter interdependence among increasingly complex or more-differentiated plants and animals. It was interdependence because in nature plants and animals appear to have been always mutually complementary and indispensable to each other.

Yet in more recent decades botanists and zoologists, although sharing the dependence upon plants, had found within their respective fields an absorbing measure of interest and challenge. Furthermore, emphasis on either field unavoidably interfered with the attainment of knowledge and appreciation with respect to life comprehensively viewed as the charming mediator of the organic world both now and throughout the ages. Such an interference was of little concern during the early stages of the botanical and zoological sciences when the emphasis was placed upon collecting, identifying and classifying organisms. It became of great concern, however, when attention became directed to patterns of behavior, to physiology, to protoplasmic metabolism, the more so because with the single exception of chlorophyll-mediated photosynthesis the magical attribute called life apparently was sparked by the same protoplasmic dynamics in plants and animals. Moreover, these developments brought about what might be termed retrospective insight into the evolutionary aspect of the interdependence of plants and animals. Long before the dawn of recorded history there was the dawn of humans and the contemporary plants and animals of the environment which conditioned their evolution in the patterns which eventually survived. Long before the dawn of recorded history there was a domestication of plants and animals by humans, primarily to facilitate ease of survival. Relatively recent scientific studies of still-primitive aborigines have neatly emphasized that long before the dawn of recorded history the more nimble-witted humans dominated and exploited other humans through alleged consort with imagined spirits purported to be capable of exerting supernatural forces: basically the domestication of plants and animals included humans. This development became very important in relation to the background of biology, since both plants and animals entered into the repertoire of witch doctors. The gradual emergence of medicines as effective curatives practically forced attention to specific forms and thus engineered the early descriptive beginnings of botany and zoology. The development also had a great influence upon the attitudes made evident in early botanical and zoological works, since there were notions that humans were the special handiwork of a supernatural creator obsessed with a whimsical indulgence in symbolism. For a time early botanists and zoologists sought clues as to the potential medicinal uses of lower organisms: both sciences were involved in these detection enterprises. From the standpoint of the historical backgrounds of biological science, however, such notions as a supernatural creator, special creation for humans and symbolism constituted opiates to pertinent intellectual activity and triumph over ignorance through the experimental methods which initiated the modern age of science.

The development of microscopes brought about the first recognition of cells as the fundamental units of life in plants and animals and thus supplied a natural basis for biology as a science. However, for a satisfying understanding of these units it was advisable to begin studies with one-celled plants and animals and to procede from such material to the study of plants and animals comprised of cells aggregated to form increasingly differentiated complex structures in which the fundamental unit cells were marvelously integrated. Basically this meant a study of natural evolution, a study which logically and inevitably led directly into the origins of concepts of supernatural forces as adjuncts to the domestication, or domination and exploitation, of humans by humans. Even so there followed a progressive differentiation of botanical and zoological interests, naturally accompanied by a neglect of shared and complementary interests.

The development of appraisals such as that represented by Darwin's Origin of Species might well have exerted a wholesome unifying influence on botanists and zoologists had it not been disconcerting to an influential group of humans which in a most natural manner, as self-appointed shepherds, had taken over with appropriate embellishments the prerogatives of the for-the-most-part remotely antecedent witch doctors. Unfortunate, also, was the emphasis which had been placed on the survival of the fittest. As phrased, the survival of the fittest carried an implication of warfare readily subject to correlation with the history of tribes and nations. Natural evolution, as will become apparent later, had many more charming behavior patterns than that denoted by the term "warfare".

Historically it was perhaps sufficiently accurate to conclude that vitalism - the explanation of mysterious activities among living organisms by recourse to imagined supernatural forces—died a lingering death during the closing years of the 19th century. It was not mourned by botanists or zoologists, even though its death ultimately had very far-reaching effects on the two disciplines. For one thing, the death of vitalism meant that all of the diverse processes recognizable in plants and animals simply had to be attributable to natural forces, ostensibly in the areas of physics and chemistry. This clearly represented a world of challenge, and because of this challenge it was not as clearly obvious that for another thing the death of vitalism hastened the eventual emergence of biology as a science. Still less obvious at the time was the practical and inevitable goal of the emergent biological science: to orient humans within a fantastic and most charming natural world. Extending the frontiers of knowledge eventually became subject to recognition not only as a progressive step in natural evolution but also as a step rich in potentialities for the enrichment of civilization and the ultimate emergence of an enduring society of really-intelligent humans.

Plato put it this way: the world's humans were in a deep abyss of darkness, meaning ignorance, and were forever building little ladders in endeavors to climb out of the dark abyss into the light, meaning knowledge. That allegory was given many centuries ago, but humans are still busy building ladders and striving to obtain more knowledge.

With respect to the attainment of knowledge concerning modes of initiation for the organic world some relatively recent developments have been of substantial promise. One of these was the attainment of evidence that amino acids—substances intimately involved in the synthesis and degradation of proteins—were subject to synthesis from a variety of mixtures of simple gases through such natural agencies as heat and electricity. Allied with this development has been an extensive study of aggregates built from the derived amino acids. Some of these aggregates have formed buds and thus have exhibited the behavior patterns for the reproductive potentialities of some of the simpler types of plants and animals. Insofar as the composition of the aggregates was concerned there was spontaneous generation and self perpetuation, both attributable to natural forces. The results necessitated a revision of popular and traditional attitudes in regard to spontaneous generation. Another relatively recent development of substantial promise was the correlation of carbon-dating and archeological discovery which constituted evidence that humanoid creatures had inhabited the earth for at least two million years. In the geological time scale organic evolution took a far longer period in attaining the conquest of the earth or aerial environment than in attaining the subsequent differentiation of the land-adapted or air-adapted organisms. In all humility, therefore, the aspiring biologist immediately was made aware of the fact that even the least differentiated organisms were not simple in the sense that their behavior patterns were understood. There was no evidence that any direct ladder leading from darkness to light was under construction, but at least there was ample evidence that research yielded light. The highway of travel for evolution had been a very long one and the curiosity of humans had only recently begun to seeth and become intellectually dynamic, retrospectively and introspectively. Quite naturally the major concern of humans had been food, and how to make a living is still a primary concern. Increasingly, however, it has become apparent in our highly diversified society how to make a living, how to enjoy making it, and how to enjoy the living itself, are three jolly companion themes. Impulsively there arose the thought that ignorance was the snorting, fire-breathing dragon to be slain. Ignorance concerning evolution seemed to be an appropriate victim.

Evolution turned out to be finest drama. Humans were not only a part of nature: they were a rousing part of it. There were the near surviving relatives: apes, monkeys, orangutans, baboons, gorillas and kin. For simplicity they could all be designated as "apes." Some of the apes had the habit of eating three meals a day. Many humans have delighted to ape the apes in their eating habits and there are many undernourished humans today who would like to adopt the habit. Some of the apes chatter interminably while others are silent introverts. Civilization witnessed congregations on the one hand, the denizens of convents and monasteries on the other hand. Some of the apes have been known to tease, or to make faces, or to play tricks on their companions, or to pound their chests and strut pompously. Humans have been known to do these things. Some apes are solicitous parents and other are not. Some apes are fighters and others cower. Some apes are dominated and exploited by humans, are taught to ride ponies and bicycles, to skate, to drink from bottles. Obviously some apes are constrained to obey commands given by human relatives who also ride ponies and bicycles, skate, smoke and drink from bottles.

The more one knows about nature the prouder one becomes at being a part of it. It is one thing to tend sheep on a starlit night and become lyrical in psychologically-comforting songs born in ignorance from undisciplined imagination. It is quite another thing to exercise the intellect — nature's superlative achievement in evolution — in conjunction with the experimental methods of science to obtain far more satisfying conquests on the frontiers of knowledge. To deprecate thought is to turn backward on the trail that has led organisms into the glory of being human.

Biological science is more than a reunion of botanical and zoological sciences. As separate sciences each had followed similar routines of description., classification, inter-relationship and physiology. Biology as the life science is much more demanding because it not only seeks a more intimate knowledge of nature: it has an allied responsibility for integrating humans with nature. This includes humans not only lacking an adequate familiarity with burgeoning knowledge concerning nature but also humans indoctrinated with fictitious concepts. Throughout the history of mankind many humans have been endeavoring to integrate themselves and other humans not with nature but with imagined supernatural forces. The involved viewpoints became traditional and habit-forming. Habit is an energy-saving device of nature and one may not be overly scornful of it, but biology bespeaks a knowledge of nature, which in turn bespeaks an abandonment of notions regarding the supernatural. Fortunately for humans nature supplies far more satisfying patterns for living than those dreamed up in the ignorance of past ages as the contributions of imagined supernatural gods. When millions of people comprise a society one presently expects and finds innumerable incompatibilities and dissensions. When millions of individual living cells comprise an animal or a plant one finds, apart from ill health, peaceful coordination for the good of each individual cell and for the good of all cells as a group. Biology as a science has a future rich with opportunities and responsibilities.

CHAPTER 2.   ORIENTATION

During the years of the present century the dynamic revolution which created a world of science has included changes in fundamental concepts having an intimate relation to biology.   At the outset of the century it was not usual to find that the supernatural in religion had its counterpart vitalistic explanations of mysterious behavior patterns in plants and animals.   Although science would have no part in these explanations it had little of assurance to offer. On the one hand there was a recognition of a responsibility for explaining behavior patterns in terms of natural forces.   On the other hand, botanists and zoologists ostensibly were dependent upon physics and chemistry, and in this critical era there was not even a satisfactory theory of solutions.   There was appreciable recognition of attractive forces operating between dissolved substances and their aqueous solvent.   There was appreciable recognition of freedom among solute units in an aqueous medium as documented by electrtcal conductivity measurements and embodied in the Kohbrausch Law of the Independent Migration of Ions.   Scientists confronted by the complexities built into protoplasm in the incomprehensible eons of organic evolution might well have shared some of the hopeless frustration of cattle seeking to comprehend automobiles except for that precious spark of superior wit possessed by humans.

Ameliorating circumstances followed.   In 1920 there came a recognition of the fact that the length of the periods of light and darkness often conditioned certain behavior patterns in plants and animals.   In 1928 there came a recognition of the fact that chemical substances also were factors potentially conditioning certain behavior patterns in plants and animals.   These two developments— photoperiodism and growth substances—were widely acclaimed and had a dramatic impact upon biology.   Here were forces of nature operative alike within the older disciplines of botany and zoology, and these forces facilitated the emergence of biology as a science. In retrospect, botany and zoology appeared to have developed with somewhat superficial appraisals embodying no adequate appreciation of protoplasm as a common denominator of life. Yet gradually the structure and physiology of protoplasm emerged as the basic features of the life science termed biology.

In 1932 a description of periodic hydrational potentiality was published, but some of its involved principles were unconventional and documentation was fragmentary. In 1964, however, this description was published with convincing documentation in the book "Behavior Patterns of Hydration". In this book it was made evident that the behavior patterns of hydration were the result of forces subject to characterization as both physical and chemical. It became obvious also that the forces were intimately and essentially involved in the mediation of protoplasmic metabolism. Hydrational potentiality thereupon became subject to appraisal as a major enabling factor in the development of the organic world, and as such it merited recognition as a basic feature of an emergent biological science.

Hydrational potentiality as represented by the description of the behavior patterns of hydration was a force limited to ions over the weight range 0 to 184,—the range represented by the 92 naturally occurring elements,—and as a force periodic with four complete periods over this weight range. It was a force maximally capable of holding 23 H_2O^- units, inversely decreasing to 1 H_2O^- unit to complete a period. The force appeared to be essential to solubility in water, since in any aqueous solution hydrated ions were present. However, there was no clear relationship made evident between hydrational potentiality and solubility: some solutes were soluble up to the point at which the hydrational potentialities of their ions were satisfied and no further, while some solutes yielded ions capable of sharing the H_2O^- units held in hydration and entering into aqueous liquids characterized by an absence of free aqueous solvent.

With respect to the indicated behavior of hydrated solute ions in solutions under electrical stress it was noted that under such conditions the hydrated H^+ hydrogen ion commonly became anhydrous, whereas most other hydrated elememt ions when in adequate solvent retained H_2O^- units to the full extent of their hydrational potentialities. These different behavior patterns made it clear that hydrational potentialities could be changed or even lost under electrical stress, and since hydrational potentiality was an attribute limited to and characteristic of specified ions the change or loss of hydrational potentiality carried with it an implication of the loss or gain of an electron and a consequent attainment of a different state. This meant that any electron changes among solutes would modify the hydrational potentialities of the involved solute ions,—a matter of special interest in relation to protoplasmic metabolism.

As projected from the foregoing considerations the natural assumption would be that any substance soluble in water would yield ions and that at least some of the ions present in true aqueous solutions would be hydrated to the extent prescribed by the description of periodic hydrational potentiality. It was to be recognized that an ability to function in the transfer of electricity was not an essential accompaniment of the ionized state, since non-polar solute ions had been evidenced as having become hydrated. This appeared to be a matter of some importance inasmuch as in a number of instances the saturation of water with a specific gas had failed to influence the electrical conductivity of the solution. There was thus the implication that under such conditions the dissolved gases were present as non-polar ions—an implication which was of assistance in the interpretation of gas-solubility data, as may be noted in subsequent chapters. For many years it had been recognized that the presence of dissolved salts and sugars in water reduced the amount of a specific gas which could be dissolved in a given volume of liquid, or conversely, that dissolved gases reduced the solubility of salts and sugars in a given volume of liquid. This situation carried the definite implication that hydration took place in salts, sugars and gases dissolved in water and that the indicated relationship involved competitive removal of H_2O units from the original aqueous solvent. As a corollary there was the further implication that in water the H_2O^- units were held by rather tenuous but definite bondages. There was a further suggestion that the nature of these bondages and the sub-ionic atom or atoms to which they were appended was a promising subject for further research. Water was readily evidenced as a liquid capable of supplying ions with the H_2O^- formula. Furthermore, hydrational bondage became subject to appraisal as a major factor in the mediation of viscosity in cytoplasm, in this respect serving organisms in a manner suggestive of the the role of oil in many machines.

The description of periodic hydrational potentiality as given and documented in the book "Behavior Patterns of Hydration" established a mathematical background for new approaches to a more intimate understanding of nature for biological science. Basically it gave an improved interpretation of aqueous solutions. It supplied evidence that incident to aqueous solution there was a transfer of H_2O^- units from solvent to solute, a transfer which made the solute concentration greater than that conventionally assigned in the absence of a recognition of hydration. This development often was involved in the interpretation of the data of the electrical conductivity of aqueous solutions-data which had contributed appreciably to the development of the still-contemporary but untenable theory of incomplete dissociation. It supplied evidence that in numerous solutions when the solute concentration became so great that no free solvent was available for the complete hydration of all ions they were able to share their prescribed complements of hydrational H_2O^- units. Under these circumstances an aqueous liquid which was not a solution was comprised of hydrated ions whose shared H_2O^- units effected a hydrational bondage. This situation also often was involved in the interpretation of data of the electrical conductivity of aqueous liquids which, in the absence of a knowledge of hydration, had been considered to be aqueous solutions. This situation also contributed to the development of the theory of incomplete dissociation. The discovery of the relationship of hydration to solute density, described in the preceding book, made it possible to successfully calculate the specific gravity of aqueous solutions-calculations impossible in the absence of a valid description of hydrational potentiality. This development prompted a study of osmosis and eventually this process, so important throughout the evolution of plants and animals, became a valuable tool in further research. There followed evidence that osmotic membranes possessed potentialities for the dissociation of solute radicals which were greater than the corresponding potentialities of water. This unexpected development suggested for osmotically-active membranes in plants and animals a hitherto unappreciated role in nutrition.

It was obvious that the description of periodic hydrational potentiality had supplied a new and promising basis for research in biological science. On the other hand, it also had supplied a basic order for the chemical elements which was in conflict with the interpretations of contemporary chemical science. There was a natural explanation for this situation,—an explanation which might or might not be satisfying to those concerned. In 1816 Prout in England somewhat timidly ventured the viewpoint that all of the chemical elements had been built up with the basic element hydrogen. He was led to do so from the results obtained in studies of the behavior patterns of gases. Prevailing at the time was the concept that atoms were solid balls of the utmost minuteness, on which account it seemed axiomatic that such a molecular weight as 32 for biatomic oxygen would yield a weight of 16 for atomic oxygen. Atomic oxygen in the neutral state is unstable and has never been isolated or weighed. For the most part the gravimetric research pertaining to the atomic weights of the elements have involved solids, and to an appreciable extent the contemporary conventional atomic weight values of chemical science represent averages sought under the concept that each element, irrespective of its state, should have one and only one weight value. There is convincing evidence that in the solid state atoms commonly exist in a variety of forms. On the other hand, in the aqueous solutions considered in conjunction with studies of the behavior patterns of hydration the atoms were evidenced as independent in aqueous solution and thus in a state analogous to that characterizing gases. It was natural, therefore, that the behavior patterns of atomic solutes should evidence uniform integral weight values consistent with the viewpoint of Prout. If there is any bigotry in chemical science a major moiety is probably centered around the assumed integrity of the conventional series of irregular, fractional, incongruous and incorrect values for the atomic weights of the elements. The validity of the description of periodic hydrational potentiality and its four corollaries defining categories of weight was amply documented in the mentioned preceding volume concerned with the behavior patterns of hydration. Emergent biological science thus has the opportunity to progress with substantial confidence in the basic order characterizing the atoms of the antecedent inorganic world as well as the organic world of its special interest and concern.

The basic order represented in the preceding volume was accompanied by an evidenced involvement of whole numbers of H_2O^- hydrational units for solute ions, and it was the attributes of these solute hydrated ions which made possible the validation. Hydrational potentialities have a challenging and as yet largely unexplored relationship to the inorganic world, although in a somewhat groping manner the importance of hydration has been recognized in conventional inorganic chemistry. Hydrational potentialities have an even more intimate and challenging relationship to the organic world. The preceding volume contained the statement that in substantial measure hydration supplied an explanation of the benefaction of rain to a plant, of a drink of water to an animal. This was a true statement, but its truth was not made particularly obvious in a book designed to document the validity of a specific description of periodic hydrational potentiality. The present volume has been designed as an assemblage of personal ideas, observations and research results dealing more directly with matters of biological interest.

CHAPTER 3.   CONCERNING PROTOPLASM.

A major contemporary research tool in biology is the electron microscope, and quite commonly this research involves studies of protoplasm. Following the procurement of photographs representing very extensive magnifications there has come the matter of interpretation, and in this matter there often has arisen opportunity, temptation and indulgence involving speculation. Speculation is a delightful adjunct to research tending to emphasize the individuality, perspicacity and resourcefulness of the investigator. Biology began describing protoplasm as a highly viscous blob of stuff, and where biology will end with its description of protoplasm no one would venture to predict. In the meantime, however, some notes regarding hydration in relation to protoplasm may not be entirely out of order.

Protoplasm as a living metabolically-active viscous blob of stuff contains hydrated ions, and the viscosity appears attributable to hydrational bondages. These bondages may involve the simple sharing of involved hydrational H_2O^- units and may also involve a substitution of hydrational bondages for chemical electron bondages, as in gelation. It has been made evident that ions may enter into chemical combination either with a retention of their hydrational units, or with a loss of these units and the sharing of electronic fields in the anhydrous state. This versatility may be projected as a factor in the adaptability of protoplasm. In Chapter 16 it will be indicated that gelation represented as a collective attribute of pectate radicals involves a progressive and extensive replacement which insures a retention of viscosity, but it is of interest that throughout their evolution plants have retained an ability to synthesize pectate radicals. Any degree of substitution of hydrational bondages for chemical bondages naturally involves access to hydrational H_2O^- units, - a consideration which is consistent with the more common incidence of gelation in organisms of the water habitat.

A common feature of protoplasm is the inclusion of one or more vacuoles. In the preceding book dealing with the behavior patterns of hydration several chapters were devoted to osmotic activities and in the studies there reported it became quite clear that osmotic membranes had appreciably greater dissociation abilities than water. On this account vacuoles may be interpreted as capable of making available solute element-ions which otherwise would remain unavailable as constituents of solute radicals. Since vacuoles were of such common occurrence-though some were so minute as to escape notice-there was the suggestion that their primary function was in the service of nutrition through the osmotic release of atomic ions.

Within a healthy metabolically active organism the cells commonly are turgid. In mature cells turgidity has appeared to be maintained by the hydrational potentialities of the solute ions present in relation to the availability of the supply of H_2O^- units, since hydration increases the volume of the ions. Chapter 18 is devoted to a consideration of the relation of hydration to turgor. In young expanding plant protoplasts the extent of turgidity has appeared to be conditioned by the extensibility of the peripheral cell walls as well as by the hydrational potentialities of the ions present and the availability of a supply of H_2O^- units. Plasmolysis has been attributed to a withdrawl of H_2O^- units from the cell sap, bnt ordinarily these units appear to have been removed from an aqueous solvent rather than from hydrated ions. The H_2O^- units held by individual solute ions collectively may comprise all or a portion of what is termed "bound water". When H_2O^- units are shared among hydrated ions they also collectively may constitute "bound water". Solute ions have different and characteristic hydrational potentialities conditioned by weight. These different potentialities result in an unequal distribution of H_2O^- units among individuals whether independent or held in hydrational bondage by an insufficiency of free aqueous solvent.

The description of protoplasm as a viscous blob of stuff more appropriately is applicable to the simplest one-celled organisms. The situation changes with evolution. In some living plant protoplasts there is an obvious differentiation into a viscous mass termed cytoplasm, a spheroid condensed usually colorless mass within the cytoplasm termed a nucleus, and one or more vacuoles commonly liquid-filled. The cytoplasm also may contain one or more pigmented bodies termed chloroplasts. Under the conditions indicated there often takes place a more or less circulatory movement of granules within the cytoplasm. The chloroplasts exhibit potentialities for independent movement, but over an indefinitely extended time make periodic trips to the nucleus. The rhythmic nature of cytoplasmic movements has been attributed to the action of various factors. Often it has appeared likely that they were primarily attributable to the removal of H_2O^- units from alveolar tissues through the dehydrating action of atmospheric ions at moist interfaces, an action which results in the hydration of the involved atmospheric ions. In contrast the more subtle rhythmic movement of the chloroplasts to and away from the region of the nucleus has appeared to relate to the dispersal of nuclear exudates of a stimulatory nature, the movement away from the nucleus having been noted as less sluggish than the movement toward the nucleus. Under conditions of unusual stress all chloroplasts cluster around the nucleus in a manner analogous to the behavior of bacteria in the development of legume nodules. Such patterns of behavior indirectly suggest a fixation of atmospheric nitrogen by cell nuclei. Metabolic nitrogen will be considered further in a subsequent chapter.

To a medical student a study of blue-geen algae is most helpful, since their forms and physiology are analogous to those of bacteria and they are larger and hence much easier to see. To a biology student a study of blue-green algae is especially helpful, since they afford insight into some of the basic interrelationships of protoplasts. For example, tn relation to the matter of inheritance there are considerations more readily understandable than those invested in the chromosomes of higher plants and animals. In several one-celled forms a protoplast appears to divide through a median constriction into two separate protoplasts inside a common gelatinous envelope. Then each protoplasts develops an individual gelatinous envelope. These protoplast may divide in like manner within their respective gelatinous envelopes. It follows that the mother envelope persists, the grandmother envelope persists, the great grandmother envelope persists,—and depending upon the specific generation there is one to many gelatinous envelopes at birth. These envelopes may have advantages - such as tending to restrain bacteria or other potentially-destructive organisms - and may have disadvantages - such as interfering with the inward passage of light, oxygen and nutrients and with the outward passage of carbon dioxide and metabolic wastes. Certain it is that in many instances the size of the ultimate colony is limited and that new colonies are initiated only following the disruption of the old colonial structure. A gelatinous wall, therefore, may be appraised as an inheritance, and there may be variations and limitations associated with this inheritance. Within the same colonies also there are other aspects of development. With a surfeit of protoplasts in a single colony those on the periphery have the more ready access to light, oxygen and nutrients. Those in the center might be appraised as underprivileged. Some of the center plants die—and their death releases substances to their associates, which may thereupon become reduced in size, become refractive, become gonidia and possess an unusual ability to withstand environmental conditions otherwise lethal. Following the disruption of the weakened colony these gonidia may disperse, germinate and establish new colonies. The important point here seems to be that the development of conditions unfavorable for vegetative growth appeared to promote the modifications which eventually surmounted the unfavorable condtions. Vegetative growth involved hydration. Gonidia formation involved dehydration and and allied resistance to environmental conditions which were inimical to the persistence of the organisms in the vegative hydrated state. Thus among the very simplest plants hydration and dehydration have respective orders of usefulness. Among higher plants one need only cite corn as as example of the incorporation of the same orders of usefulness. Up to the point of the "milk" stage of the kernels hydration has been the order of the day, but then dehydration takes place and the mature grains become resistant to environmental conditions inimical to survival in the hydrated state.

In plant physiology the evolution of dehydrated seeds as contrasted with the viviparous non-dehydrated seeds of certain tropical plants has beeu interpreted as a device to permit the extension of seed-bearing plants from the tropics into the colder latitudes. As thus appraised, hydration and dehydration have had complementary roles in furthering plants in the evolutionary project of "conquering the land".

Another aspect of hydration and protoplasm, and one more closely allied with the contemporary pattern of biological research at the molecular level, is that afforded by the writer's studies of the properties of "osmotized" versus "unosmotized" solutions. While it was obvious from the reported studies of specific gravity as an index of solute states that without resort to osmosis solubility itself denoted ionization and potential hydration, it was obvious also (1) that in some solutions not all types of ions present were hydrated (e.g. Table 36, Chap. 15 of the book "Behavior Patterns of Hydration) and (2) that in some solutions radicals were intact before osmosis (e.g. Table 3 and 4, same reference) and were dissociated after osmosis (e.g. Tables 15 and 16, Chap. 7). Collectively these developments prompted a sort of lateral or subsidiary inquiry into the behavior of some simple solutions before and after osmotization. As usual, the results obtained prompted further inquiry into the behavior of mixtures of solutions, and these results in turn directed inquiry into the behavior of expressed plant juices. The results obtained were fascinating beyond all personal anticipation. The method used was to place seriate drops of the respective liquids on clean microscope slides exposed to dust-free conditions at laboratory temperatures. After 48 hours the slides were examined under magnification. The first obvious result was the fact that hydration was correlated with filmoid deposition. This was not surprising in view of the adhesive property exhibited by a meniscus, but it was particularly emphasized when the osmotic dissociation of radicals had been involved. The startling results were those in which combinations of osmotized solutions which originally had contained intact radicals were present. Here there was a fairy wonderland of crystals. In some combinations there were islands of crystals superimposed upon a filmoid base. In others the superimposed crystals were assembled in such a way as to form dendritic patterns. There was also the suggestion that the dendritic depositions on rocks and within rocks in nature might well have involved the extrusion of osmotized solutions of hydrated atoms from the roots of plants and might not have arisen through the haphazard infiltration of simple solutions. Most surprising of all was the incidence of circular islands of crystals, each island with a common smooth circular external periphery within which every crystal was oriented centripetally in the manner of a geode.

In the examination of expressed plant juices it was noted that these were all evidenced as having contained hydrated solutes exclusively. When the study was extended to include the exudates from the stigmata of flowers it became obvious that these also contained hydrated solutes exclusively. In conjunction with these tests it was noted that the exudates from lily stigmata also contained oil. For a brief interlude this observation carried a wisp of exhilaration, since it was different and unexpected. Then in a history of technology there was noted a report that the Egyptians extracted oil from lilies at about 3000 B.C.   Collectively, however, the results, whether new or old, seemed to indicate that hydrated atomic ions had potential roles not only in the synthesis of more complex groupings as molecules but also in contributing to the developmental patterns characterizing growth. In the indicated procedures the evaporational losses which mediated the hydration of atmospheric ions naturally initiated solute concentrations which involved for certain atomic ions the presence of hydrational bondages. Previously in studies of specific gravity some solutes had evidenced a potentiality for hydrational bondage while others had evidenced an absence of such a potentiality, solubility having been limited by the individual hydrational potentialities.   There was the suggestion that these differences were involved in the exercise of specific contributions to growth. The behavior patterns of the entities termed "genes" in the final analysis must originate with atoms and their inter-relationships with each other and with their aqueous solvent.   It is no doubt a long journey from the behavior patterns of associated heterogeneous hydrated atomic ions undergoing dehydration on the surfaces of glass plates to morphogenesis in plant and animal structures,- but the indicated pathway well might be a delightful one to follow.

CHAPTER 4.   THE WATER HABITAT.

There are many reasons for concluding that life originated in ancient seas, and from an evolutionary viewpoint it seems appropriate to consider some of the probable characteristics of the original habitat.   There are various criteria for appraising these characteristics and the choice rests appreciably with the appraiser.   In the present instance the appraiser is a botanist, and the blue-green algae of today will be interpreted as surviving souvenirs of the original habitat who will display much of the morphology of their remote ancestors.   The interpretation is in keeping with the presence of gill slits - a feature of many animals of water habitats - at an early stage in the development of today's human embryos.   The blue-green algae, however, as a group represent an appreciable degree of differentiation and by implication thus represent an extensive geological period for the involved evolutionary modification and adaptation.   In this discussion it will be assumed that contemporary one-celled blue-green algae, with every cell an individual plant, are satisfactory representatives of the ancestral founders of the plant world.

The water habitat of the ancestral founders was not quite that of today's seas: the earth was younger, and hotter, and probably was enveloped in a complete spheroid sheet of water vapor.   That the earth was younger is not debatable.   The algae of today's hot springs are blue-green algae, and they are the only chlorophyll-containing plants capable of surviving and thriving at the high temperatures involved.   The blue-green algae also thrive at light intensities so low as to be incapable of sustaining any other chlorophyll-bearing plants.   The blue-green algae contain an accessory pigment which absorbs the longer wavelength radiation which is transmitted by water vapor, and only a few other plants,- the simplest of the red algae - contain this pigment.

Although speculation is involved as unavoidable it seems logical to assume that the salt content of the ancient seas was not very different from that of the present seas. It is true that increased solutes have been projected as having been added by river effluents and underwater solution, but it is also true that vast amounts of solutes have been lost to the seas through deposition, especially in land-locked embayments. The blue-green algae of today thrive best in an alkaline medium, whereas the great majority of land plants thrive best in an acid medium. In various "salt licks" of inland Louisiana.—salty springs or water holes attractive to deer and other animals—the algae commonly present in isolated pools rimmed or encrusted with salt are one-celled blue-green algae and they remain viable even after desiccation within masses of salt crystals. It may be that this tolerance represents a sort of genetic reflection of a marine origin in ancient seas no less salty than those of the present time. It could also be conjectured that the high salt content of the matrix so reduced gas solubility as to deplete oxygen for respiration and induce a metabolic dormancy. The water habitat of these "salt lick" blue-green algae is subject to great changes in salinity, and apparently they thrive equally well in fresh and salt waters. This is not surprising, since many other algae appear to adapt readily to changes in salinity. On Prince Edward Island it was noted that following high spring tides several higher types of brown and green marine algae which had become established at inland stream stations continued to survive in fresh water during the following summer, although in somewhat modified forms.

With protoplasm projected as originally born of the sea and containing hydrated ions abstracted from the sea, it became of interest to consider the densities of hydrated ions relative to water and to sea water. In the preceding volume there were nine Tables of data pertaining to the specific gravity of aqueous solutions. A wide range of solutes was involved, including most of the major solutes present in the sea. From these data the densities of each of the involved hydrated ions was readily available, and density values ranged from 1.2505 to 1.5384 relative to water. Relative to sea water the range would include unity. The gelatinous sheath materials characteristic of many algae of water habitats are substances in which to an appreciable degree hydrational bondages have become substituted for chemical bondages, on which account the densities tend to be less than that of sea water, and buoyancy is one of their contributed attributes. Naturally the question arose as to whether or not the gelatinous sheaths possessed any osmotic potentialities. In experiments designed to test the osmotic behavior of an agar matrix there was no evidence of such activity

Throughout most of the organic world life has been intimately associated with an access to oxygen. Fortunately it has not been necessary to understand with anything like precision the mechanisms involved in the essentiality of oxygen to life, but these mechanisms nevertheless comprise an outstanding challenge to biological research. For the earliest plants it commonly has been assumed that atmospheric oxygen was soluble in the surface layers of ancient seas and that this dissolved oxygen diffused into protoplasm and mediated respiration. Assuming that the projection had reasonable accuracy it still fell far short of understanding. What oxygen does for life and how it does it are what truly might be termed vital problems. One might venture that the basic features of respiration were the same in the water and aerial habitats. One might venture further that hydration was involved. Yet even the status of dissolved oxygen has remained a mystery. In conjunction with the description of periodic hydrational potentiality special consideration has been given to oxygen as a metabolite in Chapter 12.

For the earliest plants it commonly has been assumed that atmospheric carbon dioxide was soluble in the surface layers of ancient seas and that this dissolved carbon dioxide diffused into protoplasm and became involved as a source of carbon in photosynthesis. It was of interest that carbon dioxide, appraised as released incident to respiration, was much more soluble in water than either oxygen or nitrogen. It might be assumed that at the time life originated on the earth the inorganic sources such as natural vents in the earth's crust supplied atmospheric carbon dioxide, though at the present time it appears more likely that organic sources such as the respiration of plants and animals and the combustion of organic materials supply most of the carbon dioxide involved in photosynthesis. Notwithstanding the indicated assumptions regarding carbon dioxide the precise status of the gas in water has remained elusive. That its relationship to an aqueous solvent is different from that of oxygen and nitrogen has been evidenced by the fact that under appropriate pressure of the gas the solute becomes ionic with the polarity of an electrolyte. Yet when such pressure is removed the polar status is soon lost. Carbon as a metabolite has been considered in conjunction with the description of periodic hydrational potentiality in Chapter 14.

For the earliest plants it has been assumed that nitrogenous atmospheric gas, not necessarily the biatomic nitrogen gas of the earth's present atmosphere, was the source of nitrogen. It has been assumed further that the gas was soluble in the surface waters of the ancient seas and that as solute the nitrogen diffused into protoplasm and entered into metabolic syntheses. However, the precise status of solute nitrogenous gases has remained enigmatic. In retrospect it has not appeared surprising that so little has become known regarding the behavior patterns of dissolved gases when it is realized that the behavior patterns of dissolved solids as solutes also have remained largely unknown. Nitrogen as a metabolite has been considered in conjunction with the description of periodic hydrational potentiality in Chapter 11.

For the earliest plants it has been assumed that solute sulfur as a constituent of sulfate radicals and solute phosphorus as a constituent of phosphate radicals were present in the ancient seas. It could be assumed that these substances originated as volcanie gases and became dissolved in water, where their precise status remained uncertain. Sulfur and phosphorus as metabolites have been considered in conjunction with the description of periodic hydrational potentiality in Chapter 15.

For the earliest plants it has been assumed that all essential nutrient elements were present as elemental atomic solute ions or as constituents of solute radicals in the ancient seas. It has been projected that all solutes present in seawater may have potential roles in protoplasmic metabolism. Such a projection has obvious charm and challenge, but its validity must wait upon much further research. Possibly relevant to such research have been the results obtained in studies of iron as a metabolite as reported in Chapter 16.

CHAPTER 5.   SOME ADAPTATIONS TO THE WATER HABITAT

Inasmuch as life commonly has been assumed to have arisen in ancient seas it was of interest to consider some aspects of early evolutionary adaptation to the water habitat. For the most part considerations will be limited to the blue-green algae, interpreted as representing the most primitive plants.

For reasons which have been mentioned one may conclude that the earth's early temperature was higher than at present and that the earth's atmosphere at the time included a hydrosphere or complete envelope of water vapor. These two projected conditions are inter-related: because of the higher temperature there was less condensation of water vapor. Interpreted as a sort of surviving series of relics or souvenirs of ancient times, all of the contemporary blue-green algae contain a basically uniform blue pigment, called phycocyanin, which absorbs the longer wavelength red radiations of the visible spectrum. This pigment has been termed an accessory pigment: it is not essential to photosynthesis, but it absorbs radiation which the more readily passes through water vapor and which in its absence might be lost to the organism. At the very least some of the absorbed energy is converted into heat which might be projected as facilitating the production of photosynthate. This pigment also occurs in some of the simpler types of red algae, commonly appraised as the group next above the blue-green algae in evolutionary time and differentiation. It is of interest also that the accessory red pigment characteristic of the red algae occurs as well in some of the more complex types of blue-green algae. The "higher" blue-green algae and the "lower" red algae thus include plants with two types of accessory pigments. The red pigment has been interpreted as representing an adaptation for the absorption of the blue light tending to be exclusive in water of some depth.

As noted, the presence of a complete water vapor envelope about the earth was correlated speculatively with a higher temperature for the earth. The contemporary algae typical of hot springs are blue-green algae, and some of these thrive at temperatures of the order of 70°C—too hot for human hands. Consistent with the projected earth conditions at the time plant life originated, one might consi-der the contemporary blue-green algae which thrive in water at more ordinary temperatures as plants which had become adapted to water at the cooler temperatures.

Another important aspect of the blue-green algae is their present ability to thrive at light intensities too low to sustain photosynthesis in most higher plants. A light intensity of the order of 50-foot candles is adequate for the production of photosynthate in many species, whereas a light intensity of the order of 400-foot candles is minimal for many species of higher plants. It is of interest that this persistent ability of blue-green algae to thrive at light intensi-ties of a low order is consistent with the projected presence of an ancestral cloud envelope about the earth. In the waters of the Gulf Coast carrying sediments from the Mississippi river the ability of certain blue-green algae to thrive in light at low intensities has important ecological and economic aspects. The amount of visible radiation able to penetrate the sediment-carrying waters varies, but commonly it is very small. Yet some attached forms of blue-green algae are able to thrive on substrates such as shells, and these algae are extensively and repeatedly cropped by larval shrimp, often comprising their principal sustenance. In substantial measure the blue-green algae of these waters play an essential role in the develop-ment of the shrimp and the industry they sustain. More than that, they evidence adaptation to substrates and thus suggest that their remote ancestors became adapted to substrates which developed incident to the emergence of land masses from ancient seas.

The most primitive types of blue-green algae are one-celled plants, yet some of these types have developed interesting adaptations to environmental conditions in the water habitat. One of these adaptations is to water which includes high concentrations of carbon dioxide, under which conditions some discharge ions which precipitate incrustations of carbonate. Collectively such algae often have accounted for extensive limestone deposits. Yet when one-celled plants became involved in the formation of colonies by virtue of the presence of a gelatinous exudate, troubles began. There was the cutting off of light by shading and the cutting off of oxygen and nutrients by peripheral members of the colony. Motility appeared to have developed as an evolutionary innovation to circumvent the indicated handicaps. In the simplest one-celled blue-green algae motility has appeared to be restricted to germlings emergent from refractive gonidia, and the motility commonly has been noted as of short duration, of the order of a few minutes. On the other hand within the genus Oscillatoria four types of motility have been noted (1) an oscillating movement of the filament of cells, memorialized in the genus name as an outstanding characteristic (2) a forward or backward movement in the direction of the filament axis (3) a spiral turn of the filament, with or without a forward or backward movement and (4) an undulating movement characteristic of germlings moving within a gelatinous matrix. In several species of blue-green algae having filamentous structures there may be noted a modification of one or more terminal cells to provide chlorophyll-less but carotin-containing pilot cells which appear to have enhanced sensitivity to light and to have the ability to guide the forward movement of the filament either toward or away from a light source, depending upon the intensity of the light. After these pilot cells lose their chlorophyll they seem unable to sustain their activity through photosynthesis, a fact which appears to indicate that carotin has no ability to produce photosynthate. These pilot cells exhibit a conspicuous opening in the cell wall in the region of the contiguous cell, which also exhibits a similar and confluent opening. At times a passageway may be noted as containing protoplasmic material. What is more surprising is the presence of plugs or "cellulose buttons" in each cell, of a size neatly fitting the opening and usually disposed near the opening. At times these "buttons," as though by mutual agreement, are plugged in at the respective openings in the walls. In some cases there may be more than one cell of a filament adapted to the maintenance of a food supply to the pilot cell. When three support cells are involved the central cell has two openings and two "cellulose buttons." It is of interest that "openings" and "buttons to fit" are common in many species of red algae, where they are also involved in the translocation of photosynthate.

In contrast to the floating-free types of blue-green algae there are species exercising devices for attachment, at least theoretically representing the arrival of substrate surfaces incident to the emergence of land masses. In some instances the gelatinous exudate has obvious agglutinating properties. Within the genus *Nostoc* some species develop free-floating colonies while other species develop attached colonies. The most notable attachment device, especially for correlation with the red algae, is the production of a flat plate or lenticular pediment of cells adnate to a fixed surface. Following the development of such a structure filamentous outgrowths at right angles to the substrate surface are produced. This type of arrangement represents an adaptation to a swift water habitat, and is characteristic of many species of red algae. Within the group of blue-green algae the genus *Capsosira* has these two types of structures as an adaptation to swift water. It is of interest that the filamentous structures which develop at right angles or perpendicular to the substrate surface do not ordinarily evidence any phototropic sensitivity.

One of the alternative adaptations of a blue-green alga to a swift water habitat is that represented by the lichen *Ephema*. Commonly in lichens the fungus component sets the pattern of development and the contained commensal algae are inconspicuous. In *Ephema* the alga sets the pattern of development and the fungus is inconspicuous. The fungus component extends as a transparent glove covering every part of the much-branched alga. Obviously the covering lacks the ordinary function of a protection against desiccation, since the lichens are submerged. Yet the fungus has a special role: it develops a stem-like growth at the base of algal fronds and a strong rhizoidal holdfast which anchors the lichen to rocky substrates in swift water.

From the foregoing considerations one may venture that during the long geological period in which the blue-green algae were the only plants of the earth capable of maintaining themselves through photosynthesis there was an emergence of land masses from ancient seas initially darkened by a cloud envelope. From the standpoint of biology it is of interest further that the reserve carbohydrate of the blue-green algae is not the starch so common in higher plants, nor the inulin found in some higher plants, but is glycogen, the reserve carbohydrate common in animals. This product also suggests that blue-green algae originated in an early period when evolutionary differentiation of plants and animals was not as well marked as it subsequently became. There exist today organisms which contain chloroplasts and yet which also engulf food, thus combining plant and animal types of nutrition. There are also organisms whose life history includes an animal-like stage and a plant-like stage. Although these combination forms are interesting, in general biological studies have seemed more satisfying when the organisms involved were to be classed as either animals or plants, perhaps because students are so definitely animals.

CHAPTER 6. THE AERIAL HABITAT

Since humans most certainly are included among the organisms whose ancestors came up out of the ancient seas to inhabit the also emergent land masses the common associates of most humans have been the land plants and animals. The land is an excellent habitat- now that we are adapted—but it was not an easy simple thing for organisms of the ancient seas to establish themselves on land. The many trials of those early adventures may remain for the most part forever unknown. Yet although no humans stalked the earth at the time it seems rather safely speculative to hold that as the earth gradually cooled there was a diminution of the hydrosphere or envelope of water vapor. Some interpret contemporary clouds as persistent representatives of the complete envelope which formerly prevailed. In any case we have this interesting situation: the simplest types of blue-green algae, basically adapted to a water habitat, now thrive in light of low intensity and do not flourish when exposed to present day light. In contrast, most land plants thrive under present daylight conditions, but do not thrive in light of low intensity. One may venture to assume, therefore, that in general the cohquest of the land, or the adaptation to the aerial habitat, took place at a later date when the envelope of water vapor had diminished to such a degree that a substantial amount of sunlight reached the earth's surface. The aerial habitat to be conquered was one which had much more light, and sunlight included poten- tially harmful rays. After the conquest of the aerial habitat it became more readily apparent that protection from destructive solar radiation was a necessity in the aerial habitat. The high-riding pollen grains of the contemporary wind-pollinated plants have yello- wish or orange pigment which supplies the protection.

To an important extent the aerial habitat required an enhanced resistance to gravity. In the water habitat the displacement of the water by the organism conferred a factor of buoyance which was negligible in air. Linked to elevation in air there was need for resistance to wind and rain. Of even more serious nature, the aerial habitat in many latitudes and altitudes prescribed for survival a resistance to frost, with hail and snow as accessory obstacles. Certainly land or aerial habitats had unfavorable factors for organisms adapted to the water habitat.

One of the most fascinating ways to approach the "problems" confronting a prospective transition from a water habitat to an aerial habitat has involved a consideration of the major features of the land plants which were subject to appraisal as adaptations. The epidermal cells of many higher land plants and animals attest protection not only against solar radiation but also against desiccation. The dehydrating potentialities of the atmosphere of the aerial or land habitat constituted an implied factor of great importance which had to be overcome. From the description of periodic hydrational potentiality one might conclude that any hydration involved the removal of H_2O^- units by ions. In water these H_2O^- units involved in the hydration of solute ions were to be projected as held by forces and moieties which had held them. In air the dehydrating potentialities conversely meant the hydration of atmospheric ions whose nature also was not clear. Biology has some interesting and challenging problems in the area of more precise knowledge concerning the interplay of hydration and dehydration in organisms of the aerial habitat.

Another major feature of land plants and animals was the great diversity of form and physiology which developed. This feature carried the implication of a much greater variety of factors influencing development in the aerial habitat than in the water habitat. Here again speculation becomes a necessity, but one may be justified in assuming that from a number of approaches such as temperature, nutrition and light the variability within the ancient seas was not nearly as great as that within the aerial or land habitat, especially following the diverse geological history of the land masses. Perhaps it would be safe to venture that throughout the long period of organic evolution the changes in the aerial or land habitat,—some brought about by organisms - have been far greater than those within the oceans, and that these changes have mediated diversity among the evolving plants and animals. As students of nature it may be noted that for humans to be resentful of change often may be to deprecate the very pattern of behavior which introduced brainy kinfolk and early man,-and to declaim against evolution is to talk from a background of ignorance very remote from the advancing frontiers of present-day knowledge. One of the nicest things about the aerial habitat to which we have become accustomed is the autocatalytic nature of its diversity.

It is to be noted that the water habitat had an indelible effect on the organisms which came up out of the ancient seas and began the conquest of the land. Physiologically water has remained an essential constituent of all plants and animals. It is obvious that the land plants commonly have maintained roots extending into the realm of a water habitat. Among animals it is a common conceit of humans to appraise themselves as the contemporary ultimate in natural evolution,-often quite without regard to the tremendous potentialities for improvement - and yet with every breath the moistened avenue to human lungs insures that the atmospheric ions present will become hydrated by the time they reach their destination. Thus water we have always with us. Hydrational potentialities are all-pervading throughout both plant and animal physiology. Our ephemeral embryonic gill slits remind us of the continuity of evolution and the emergence from ancient seas. Our dependence on air reminds us that we have come a long way. Our ignorance reminds us that we have a long way to go. Biological science is in its infancy today, and one of the really excitings things about it is the fact that we have just begun to appreciate hydrational potentialities. As has been noted, in the preceding volume it was stated that in substantial measure these potentialities supplied an explanation of the benefaction of rain to a plant, of a drink of water to an animal. It will be a responsibility—and a delight—to increasingly implement such explanations in an intimately dynamic biological science.

CHAPTER 7. SOME ADAPTATIONS TO THE AERIAL HABITAT

In Chapter 5 the consideration given to evolutionary adaptations to the water habitat was restricted for the most part to examples within the group of simple or primitive plants known as the blue-green algae.   In the present chapter the same group of plants will be involved.   It is to be recognized that the "conquest of the land" by plants and animals led to an extensive series of differentiations in both types of organisms, including the eventual appearance of humans.   Many books could be written of the adventures embodied in the conquest.   There is space here for but a single chapter.

One of the most interesting plants of America's North Atlantic Coast is an alga which develops as an unseen blue-green stratum one to two centimeters beneath the surface of granitic intertidal rocks. Twice in every twenty-four hours the rocks are submerged in sea-water.   Twice in every twenty-four hours the rocks are exposed to the air.   Here is a fantastic approach for a marine plant with the conquest of a land habitat as an evolutionary project.   It could be stated that a solid beachhead had been established.   One may note further that the light which is able to pass through one or two centimeters of granite is indeed of low intensity.   Except for the discontinuous bits of quartz in the granite one might expect the rock layer to be completely opaque.   Possibly the rock layer often is completely opaque, for it is not unusual to find the surface of the rock coated with bacteria.   The environment represents something of a compromise.   Rains at low tide introduce fresh water.   High tides introduce seawater.   The texture of the rock places impressive restrictions on the expansive growth of cells.   They must live sedentary lives,—but for how long ? Years? Centuries?   A simple question, but an intriguing one and one not easy to answer. Apparently the algae are not particularly effective as a factor in the disintegration of the rock, as there is little or no evidence of the shearing off of thin layers from the rock surfaces. The algal cells are living: when layer-containing rock portions are placed in seawater and exposed to light of low intensities the cells divide and eventually may produce an extensive outgrowth from the original stratum. The algae cells within the rock will withstand appreciable desiccation: unwrapped portions carried in an automobile from Maine to Louisiana readily evidence regeneration when immersed in seawater at low light intensities. As to life within the subsurface rock stratum, in the natural or unnatural habitat, one is tempted to project it as representing a sentence to life imprisonment in close quarters in almost total darkness. Any excitement of adventure in adaptation to the aerial habitat for these organisms seems to involve little beyond the attainment of exclusive domain.

In sharp contrast one may turn to a consideration of the fortunes of some other members of the group of blue-green algae. One establishes beachheads in a less formidable matrix: in sand. At various places along the Gulf Coast, particularly on the landward sheltered shores of outlying islands there occurs an unusual situation. When one walks along the white sandy beach in bright sunshine and looks behind, he finds that his recent foot-prints are blue-green. This interesting phenomenon is due to the subsurface presence of countless colonies of a blue-green alga of the genus *Microcoleus*. In sunlight all of these colonies remain below the surface of the beach sand. On a cloudy day, however, the behavior of the colonies appears to depend upon the intensity of the light. Commonly numerous colonies, which have the general shape of greatly elongated spindles, and within which are numerous filaments parallel to the axis, project upright into the air for a distance of about a centimeter. The exterior common gelatinous wall then remains in this position and filaments move up into the light, stay for an indefinite period and then return to the darkened portion of the structure. Other filaments move up to positions in the light and later return. This behavior pattern obviously allows all filaments to carry on some photosynthesis on a seemingly voluntary plan of exposure to light. In the event of sunshine the entire colony will return to the moist and dark subsurface sand. On the other hand, in the event of rain the entire colony may emerge and move horizontally along on the surface of the sand. At times there are low shaded areas on the Louisiana State University campus at Baton Rouge which become green with horizontal surface colonies of Microcoleus during a rain. In this genus the presence of hydrational bondages within the individual gelatinous sheaths of the filaments facilitates freedom of movement, and yet the colony of filaments has motility. The individual filament and the colony of filaments: each exihibits a degree of adaptation to the aerial environment.

Apparently for the blue-green algae the transition from a salt water medium to a freshwater medium was not particularly troublesome. Microcystis is another genus of blue-green algae found in both marine and freshwater habitats. The plants are of special interest in Lousiana because the colonies often occur in such numbers in streams flowing into the Gulf of Mexico as to approach the bacteria-type counts of millions of cells per cubic milliliter. It is of interest further that these appear to become plasmolyzed yet not killed within oysters, and after discharge and recovery from the "milking" process remain for further contributions to oyster nutrition. This procedure appears to have been going on for some time, since the plants were identifiable in rock drill cores taken at 500 feet below ground level in southern Louisiana.

In the lower Mississippi river flood plain the early settlements of the French were along bayous which served as transportation highways. Farmlands commonly extended in linear sections at right angles to the bayous and cypress posts often were used in fencing properties. Over the years many of these cypress posts came to wear coatings which comprised silent testimony to a common pattern of conquest by blue-green algae. This pattern was an association with fungi. In general the coatings did not represent the order of compatibility, union, interdependence and composite organization of the lichens, but nevertheless it seemed quite probable that neither algae nor fungi alone could have produced the coatings.

In areas adjacent to the Mississippi river flood plain there are many aerial habitats of high humidity as attested by the abundance of Spanish moss. On certain brick walls in these habitats blue-green algae of the genus *Scytonema* long have been established, and these have formed matted coatings in which the filaments in general have tended to be parallel in the vertical adnate position. In time these coatings have extended upward beyond the point of normal exposure to rain into the area normally protected from rain by the overhang of the roof. In such extensions the algae filaments are individually wrapped by tightly contiguous coils of fungus mycelium, and the terminal cell of the upward advancing alga is neatly capped with a convex fungal sheath. Thus portions of the algae filaments might be said to be lichenized. On occasions following a rain accompanied by wind the lichenized portion of the matted coating may become well soaked. Under such conditions the terminal caps open as by a hinge and the alga extends upward growth, only to become overtaken and lichenized in a subsequent period without moisture. The situation could be interpreted as representing what might be called collusion in adaptation to drought as a potential feature of an aerial habitat. Or it could be interpreted as representing an informal and potentially ephemeral commensalism in the direction of the organic complex characterizing the lichens. Occasionally an analogous relationship between alga and fungus has been noted in the case of species of the genus *Stigonema*.

On moist soils, particularly in humid regions of the southern

States, another type of primitive lichen has been noted as common. One representative of this type involves a blue-green alga of the genus *Porphyrosiphon*, a name which may be translated to signifiy "red tube". The more one studies any alga the more interesting it becomes, but this one has features of unusual interest. There may be several filaments of cells within a common external gelatinous tube or sheath, and each filament not only has a potentiality for movement in either direction within the tube, as was noted for *Microcoleus,* but each filament extrudes gelatinous material as it moves along. On this account the common tunnel or tube in which movement takes place becomes laminated, each layer registering the passage of a filament. In older groupings the lamination may become extensive. Still more unusual is the fact that the red color which was memorialized in the genus name *Porphyrosiphon* is due to the presence of a fungus which inhabits the gelatinous sheath. Under laboratory conditions in a nutrient solution the alga may be cultured free of the fungus, under which conditions a blue-green color persists. In nature it is a primitive lichen. In southern Louisiana the combination is so common that at times appreciable areas of the soil in cane fields will appear conspicuously colored dark red. Occasionally following harrowing the red mats may become broken up into small bits. Later the development of these bits may give the soil the appearance of having been sprinkled with drops of blood.

Along roadsides and on embankments throughout many temperate regions another example of the same type of adaptation to an aerial habitat is that in which a colonial gelatinous alga of the genus *Nostoc* is associated with a colorless mycelial fungus which appears to have the relationship of a commensal parasite. It may be ventured that the presence of the fungus is physiologically related to survival on the surface of soils occasionally subject to rainfall, since no fungus-free colonies have been noted. An interesting feature of the primitive lichen is the presence of spheriod haustoria of two sizes corresponding with the internal volumes of ordinary vegetative algae cells and of larger specialized cells termed heterocysts. The larger haustoria appear to evidence for the heterocysts a dynamic metabolism sustained by the contributions of one or more adjoining vegetative cells.

The maximal conquest of the aerial habitat by blue-green algae is that attained through an inter-relationship involving organic associations between specific species of algae and fungi and designated as true lichens. Some of these lichens are able to thrive in habitats so unfavorable to other organisms as to afford them exclusive domain. Typical of such habitats are the surfaces of rocky cliffs, the surfaces of gravestones and the trunks of trees. The organic association of the two organisms in true lichens is so intimate morphologically and physiologically that each specific combination with taxonomic satisfaction has been accorded the binomial designation commonly reserved for species of plants and animals.

Occasionally analogous associations of blue-green algae with bacteria take place. On the campus of Louisiana State University at Baton Rouge two such associations have been noted. In one of these an abundance of discoid structures about the size of pennies develop on soil surfaces. These have proved to be radiate filamentous colonies of blue-green algae of the genus *Cylindrospermum*, and each colony was inclosed within a transparent cellophane-like film of bacteria. The association appeared able to withstand appreciable desiccation without injury. In a second association bacteria inhabited the gelatinous sheaths of the blue-green alga *Fischerella*, and appeared to facilitate survival on soil surfaces. Moreover, the bacteria gave off exudates which were antibiotic to numerous other species of blue-green algae and to the roots of several grasses. During the cooler parts of the year this association of algae and bacteria is typically dominant.

It would be precarious at this time to venture any atomic and molecular explanations of potential adaptability, but it is perhaps worthy of note that within countless living protoplasts there exist unsatisfied hydrational potentialities which are subject to modification. Under such conditions the loss or gain of either an electron or an H_2O^- unit would bring about a modification. The loss or gain of solutes in conjunction with translocation would modify hydrational potentialities. Yet perhaps the most significant adjunct to a potential adaptability derives from the behavior of osmotic membranes. As documented in the data of Tables 7, 9, 10, 12, 13, 15 and 16 of the preceding book "Behavior Patterns af Hydration" these membranes release hydrated ions at all membrane-liquid interfaces. With respect to an osmotically mediated increase in solution volume within an osmometer this meant that the volumetric increase was caused by half the number of anhydrous ions which had entered the membrane, the other half leaving the membrane at the opposite liquid interface. With respect to potential adaptability, however, this meant that any osmotic activity involving aggregations of protoplasts resulted in a mutual sharing of hydrated ions by all protoplasts. Moreover, such sharing embraced extrusion to a liquid environment itself. It seemed clear that in multicellular organisms the behavior patterns of osmotic membranes contributed to the potential adaptability of protoplasm, in part through the tendency toward effecting a uniform distribution of hydrated solute ions to all contiguous cells and in part through the tendency to extrude similar ions and the involved modification of the environment. In the laboratory culture of various organisms a modification of the culture medium has been a commonplace development.

CHAPTER 8.   CONCERNING CHLOROPHYLL

INTRODUCTION

For many years photosynthesis has been recognized as the process which made possible the development of the organic world.   Commonly the process has appeared to involve chlorophyll exclusively, on which account this green material has been subject to characterization as the most important pigment an earth.   Readily it was understandable that studies of photosynthesis and chlorophyll long had held prominence on the frontiers of scientific research.

With respect to chlorophyll, investigations received great impetus with the analytical work of Willstatter and colleagues in the early years of the present century.   These and subsequent similar studies were carried out in laboratories equipped with varied technical devices for performing operations designed to indicate aspects of composition or structure for the chlorophyll molecule.   The results obtained have been interpreted in accordance with contemporary practices and viewpoints.   Collectively through the years the experimental procedures resulted in the accumulation of many data and interpretations pertaining to the direction of an improved understanding of chlorophyll and photosynthesis.

For the progress made one may be very thankful, yet much information has remained to be desired in several areas.   The chlorophyll molecules have been evidenced as comprised of more than a hundred atoms, and in molecules of such complexity it has been natural that some uncertainty has attended the diagnosis of precise composition.   Still greater uncertainty has attended the projection of structural configurations.   As a consequence of these uncertainties, and superimposed upon them, there were such formidable aspects as the inter-relationships of the component atoms of chlorophyll molecules and the mechanism engineering their individual and coordinated patterns of behavior.

Because of these foregoing considerations chlorophyll became a subject of special interest following the writer's recent republication of a description of periodic hydrational potentiality in book form. This interest was not mediated directly by the description itself, since none of the four common protoplasmic plant pigments—chlorophyll a, chlorophyll b, carotin and xanthophyll—was soluble in water. The interest was engendered by the corollaries of the description which differentiated and characterized categories of atomic weight. Such corollaries presented a new and promising background from which to project a re-examination of accumulated data because the validation of these corollaries in the areas of aqueous solutions and gases has sustained also the integrity of two principles of basic significance. One of these principles held that ions entering into chemical compounds maintained therein specific and characteristic sub-ionic states—states commonly not accorded full and precise recognition in conventional chemical practice. The other principle held that ionization effected a specific change in weight,—a principle also at variance with the viewpoints and practices of contemporary chemical science. The present paper was prepared in order to document the results obtained in a re-examination of data pertaining to chlorophyll from the indicated background.

COMPOSITION FROM CHEMICAL ANALYSIS

The contributions of chemical science to mankind have been so numerous and beneficial as to have served as mute testimony to the integrity of the involved procedures. Historically the early chemical procedures with respect to plant pigments involved the use of drastically-destructive reagents and led to results which suggested that the protoplasmic pigments differed in different species of plants. With the subsequent use of organic solvents by Willstatter and colleagues, however, the analyses yielded more uniform results which suggested that the same four pigments were present in all green plants. For many years the chemical composition of these four pigments has been represented as follows:

Chlorophyll a	$C_{55} H_{72} O_5 N_4 M_g$
Chlorophyll b	$C_{55} H_{70} O_6 N_4 M_g$
Carotin	$C_{40} H_{56}$
Xanthophyll	$C_{40} H_{56} O_2$

From the formulas for composition it was evident that the difference between chlorophyll a and chlorophyll b was comprised within unit atoms of a single H_2O molecule. Although the two pigments thus appeared to be closely related, their relationship in nature was subject only to indefinite surmise. From the formulas also it was evident that the difference between carotin and xanthophyll was comprised within unit atoms of a single O_2 molecule. Any inter-relationship between the two chlorophyll pigments on the one hand and the carotin and xanthophyll pigments on the other hand remained obscure.

Collectively the compositional formulas were impressive. In the case of chlorophyll a and chlorophyll b the agreement with respect to carbon, nitrogen and magnesium was perfect, and the detection of such a number as 72 hydrogen atoms seemed to further document meticulous detail in measurement. In the case of carotin and xanthophyll the agreement with respect to carbon and hydrogen was perfect, and the detection of two atoms of oxygen in the xanthophyll molecule seemed to testify further to the sensitivity of the involved analytical procedures. Under these circumstances it was not surprising that the interpretations of the experimental data for the most part remained unchallenged. Only some such development as the discovery of periodic hydrational potentiality and the subsequent accumulation of data evidencing sub-ionic integrity for the component atoms of molecules and change in weight with ionization appeared to justify a search of alternative interpretations.

STRUCTURE FROM PATTERNS OF CHEMICAL BEHAVIOR

The reaction potentialities of the four listed protoplasmic pigments engineered the elaboration of structural formulas. On the part of chemists it was well recognized that these formulas were projections designed to represent correlations and integrations between the analytical data and the observed chemical reaction patterns of the respective pigments. In some measure these structural formulas were able to retain their original status as projections, but in some measure also the impressiveness of the analytical data and the correlated prestige of chemical science were effective in obscuring their true nature. Such developments were not restricted to appraisals of the four pigments but were equally prevalent throughout chemical science, and especially so in the areas of more extensive elaboration in organic chemistry.

For the survey at hand it was considered to be adequately illustrative to cite a structural formula for a molecule of chlorophyll *a* as projected from patterns of chemical behavior in conjunction with analytical data. The following formula in Figure 1 was taken from a contemporary taxtbook by Bonner and Galston, in which it had been adopted from a research publication by Fischer and Stern.

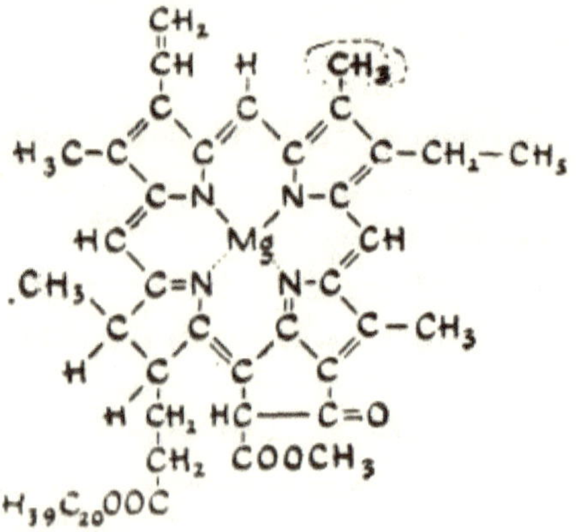

Chlorophyll a

Figure 1. A conventional contemporary structural formula for the Chlorophyll *a* molecule.

It was to be noted in the structural formula for chlorophyll *a* that except for the abbreviated segment H_{30} C_{20} the projection reflected a recognition of specific valences for the component atoms. Carbon was assigned a valence of four, nitrogen a valence of three, oxygen and magnesium a valence of two and hydrogen a valence of one. Yet it also was to be noted that the formula *failed to reflect any recognition of a specific positive or negative ionization for any component sub-ionic atom.* The failure was entirely in keeping with contemporary practice in chemical science, notwithstanding which it was appraised as most unfortunate. One might venture that the treatment extended to carbon a potentiality for covalency,—or that the importance of a maintenance of a more complete characterization for the constituent sub-ionic atoms of molecules had not been fully recognized. In any case, except as to valence the structural formula given in Figure 1 was devoid of satisfactory characterizations for the component sub-ionic atoms and thus was at variance with the abundant hydrational evidence for the maintained identity of such units in chemical compounds as given in a recent publication by the writer.

Structural formulas for commonly associated chlorophyll *b*, carotin and xanthophyll have been projected from analogous data and similar backgrounds. In each instance such formulas have evidenced a recognition of valence, but otherwise the nature of the involved sub-ionic atoms has not been indicated. In each instance, moreover, the projected structural formulas have been similarly devoid of symmetry.

CONVENTIONAL APPRAISAL

It was to be recognized that any contemporary structural formula for chlorophyll *a*,—as well as for any associated protoplasmic pigment—arose as a projection based in part on composition as evidenced by chemical analysis and in part on atomic inter-relationships as evidenced by chemical reactions. It was to be recognized further that although a structural formula of this nature was a projection, any extensive use of such a formula had a definite tendency to impart to it an aura of validity. Collectively the indicated origin and the pattern of use often have created a situation unfavorable for the impartical consideration of alternative viewpoints.

With respect to the structural formula for chlorophyll *a* it has seemed quite likely that although differences of opinion might exist among biochemists the general arrangement of the constituent atoms would be in the pattern given in Figure 1. For this reason it also has seemed quite likely that any drastic departure from such a pattern would be highly provocative to those particularly interested in the subject. Unavoidably the introduction of alternative viewpoints has seemed accompanied by implied criticism of these viewpoints they might or might not be destined to supplant.

Historically it has appeared that before the classical researches of Willstatter chemical analyses involved the rather complete disorganization of the protoplasmic pigments. The use of organic solvents by Willstatter initiated an era during which chemical analyses involved a far less drastic disorganization of these pigments. Nevertheless, such structural formulas as the one given for chlorophyll *a*, in the opinion of the writer and for reasons made apparent in subsequent pages, have evolved from chemical reactions of impairment.

PERIODIC HYDRATIONAL POTENTIALITY

Inasmuch as the four protoplasmic pigments under discussion were insoluble in water it seemed natural to assume that hydrational potentiality had no immediate relationship to the composition or structure of the pigment molecules. The assumption seemed valid except for one important item: the description of periodic hydrational potentiality included the recognition of categories of atomic weight, and contemporary conventional chemical science had evolved in the absence of any such recognition. The description was first put forward in the journal of the Washington Academy of Sciences in 1932, but a comprehensive documentation of evidence validating the description did not follow until it appeared in book form in 1963.

With respect to the composition and structure of the four indicated pigments the important feature of the description was its third corollary characterizing the combining weight of ions. It was to be recognized that for many years chemical science, often with an obvious lack of assurance, had entertained a precept which characterized combining weight as a value obtained by dividing an atomic weight by a valence. Under this precept the atomic weight of an element was identical with a combining weight for the element in an ionic state when the valence was one. When the valence of the ion was more than one, however, the value derived as a combining weight became abstruse and ambiguous, having a practical relationship to procedures anticipating reactions, but devoid of a self-contained significance.

In contrast to the indicated precept of chemical science the combining weight of an ion as prescribed by the description of periodic hydrational potentiality was obtained by adding the respective weights in the neutral and ionic states and dividing the sum by 2. The derivation was the same whether the involved unit was an atom or a molecule, an element ion or a radical. Since neither atoms nor molecules evidenced any hydrational potentiality while in the neutral state their weight values in this state had a basic but only indirect relation to the behavior patterns of solutes in an aqueous medium. Moreover, because hydration was restricted to ions the combining weight values of solute ions were of subordinate significance in relation to the consideration of phenomena involving the attributes of solutions. Yet in relation to the interpretation of data involving molecules insoluble in water the prescribed combining weight values were subject to recognition as attributes having a major order of importance in the interpretation of observational data.

ATTRIBUTES OF SUB-IONIC ATOMS

In the absence of satisfying interpretations of the protoplasmic pigments there was a natural resort to empirical projections. These were initiated from the implied precept that the organized component atoms of all molecules entered into chemical combinations as specific ions and maintained their integrity as ions while in the combined state Under this precept the component atoms of the molecules were subject to differential characterization as sub-ionic atoms, and as such the contribution of each component atom to the weight of the molecule as an integrated unit was its *combining weight*.

As thus appraised it was obvious that the specific nature of the ion, positive or negative in addition to valence, was a matter of potentially great importance in the projection of structural chemical formulas. This was especially the situation with respect to molecules containing carbon, since this element often had been conceived and treated as having possessed the ability to form covalent ions simultaneously amphoteric. As prescribed by the description of hydration no element ions possessed such an attribute: their ionization was either positive or negative, never part positive and part negative.

To what now seemed an unfortunate extent the specific character of sub-ionic atoms in molecules had failed to receive attention in chemical science, and a similar situation prevailed with respect to the specific change in weight with ionization. As has been indicated in the 1963 documentation of evidence validating the description of periodic hydrational potentiality, these two oversights were directly involved in the meticulous elaboration of the contemporary system of irregular fractional atomic weight values.

The relation of these two oversights to the interpretation of analytical data pertaining to the four common protoplasmic plant pigments became apparent incidental to the empirical projection of structural formulas consistent with the indicated principles associated with the description of periodic hydrational potentiality and its corollaries differentiating categories of weight.

CHLOROPHYLLS *a* AND *b*: STRUCTURE AND COMPOSITION

Inasmuch as chlorophyll *a* without chlorophyll *b* was present in the bluegreen algae it was ventured that the latter pigment had developed as an accessory rather than as an essential adjunct to photosynthetic potentiality. On this account attention at first was centered on the structure of chlorophyll *a*. Without apology the procedures were purely empirical and conjectural, but they had the distinct advantage of projection from a background which assigned a maintained identity to sub-ionic atoms deprived of covalency. The procedures clearly would have had no direction beyond whimsy, however, in the absence of the efforts of analytical chemists. Yet with the experimental results as a sort of compass, something closely akin to progress seemed to follow. A contrasting and striking feature of the developments was the endowment of structural formulas with a degree of radiate symmetry entirely lacking in the conventional structural formulas. Of necessity the structural formula presented here represented symmetry in a single plane, but it became obvious that spheroid conformations and molecular linkages would supply intriguing challenges for some time to come. For the present the consideration of a molecule appraised as an aggregation of integrated sub-ionic atoms in a single plane was quite enough.

The molecule of chlorophyll *a* as thus appraised and projected has been represented in Figure 2.

A comparison of the projected molecule given in Figure 2 with the projected molecule given in Figure 1 revealed some important similarities and some important differences. With respect to composition the numbers of atoms of hydrogen, oxygen, nitrogen and magnesium were identical,—a fact which rightly or wrongly

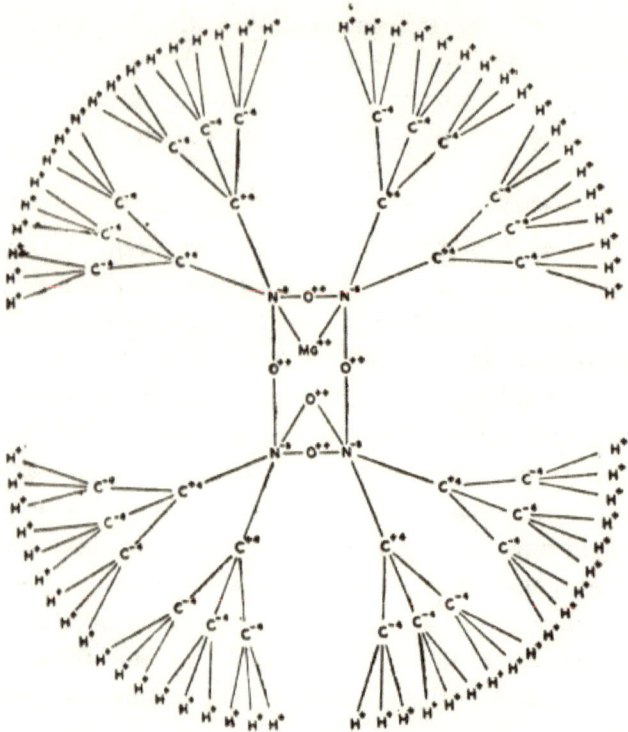

Figure 2. Projected structural formula for Chlorophyll *a*. Composition: $C_{32} \ H_{72} \ O_5 \ N_4 \ Mg$

seemed to impart a substantial measure of confidence not only to the work of the analytical chemists but also to the symmetrical projection. It was to be noted that the structure and composition of the central portion or so-called nucleus of the molecule was quite different in the two projectsons. In the symmetrical projection four oxygen sub-sonic atoms were in positions of sub-sidiary resonance and one was in a position of major susceptibility to resonance. The single sub-ionic atom of magnesium was represented as occupying also a position of major susceptibility to resonance in contrast to the more conventional structural pattern in which it was represented as subject to unstable interchange linkage.

The major ostensibly incompatible feature of the projected symmetrical molecule was its content of carbon. In the conventional molecule projected from analytical data and reaction patterns there were 55 atoms of carbon, whereas in the symmetrical molecule there were 32 atoms of carbon. This incompatibility could not be ignored: it had to be explained.

To the unabashed surprise and delight of the writer the analytical data of chemical science for the composition of chlorophyll a supplied unwitting but outstandingly impressive evidence of the validity of the precise change in weight with ionization which had been embodied in the description of periodic hydrational potentiality. Heretofore the evidence as documented (1963) had involved relatively minor changes in the behavior patterns of gases, solutes and solids. The evidence had been consistent, extensive and impressive, but it had not included any such degree of modification as that involved in the transition of carbon atoms from a status as quadrivalent negative ions to a status as quadrivalent positive ions.

As prescribed by the description of periodic hydrational potentiality the combining weight of a quadrivalent negative carbon ion, C^{-4}, was 8, whereas the combining weight of a quadrivalent positive carbon ion, C^{+4}, was 16. In the chemical analysis of chlorophyll a the carbon presumably was driven off by heat and weighed as the dioxide gas or the carbonate solid, in both of which states the carbon sub-ionic atoms were indicated as quadrivalent and positive. This meant that each of the 24 C^{-4} sub-ionic carbon atoms indicated in Figure 2 was weighed in such a manner as to have been accorded a weight which was twice its weight in the original chlorophyll a molecule. Under such circumstances it was natural that the number of such atoms was evaluated as double the actual number. The 8 quadrivalent positive sub-ionic atoms of carbon in the original chlorophyll a molecule had the same weight values in the resultant compounds following denaturalization. Thus the total number appraised from gravimetric data in the absence of a recognition of change in weight with ionization should have been subject to calculation as $24 + 24 + 8$, or 56. In the conventional compositional formula the number of carbon atoms was to be recognized as 55. At this point the order of agreement was perhaps well enongh, but speculation was easy and the analyses had been laborious. Empirically projected there was the suggestion that in the analytical procedures the nuclei of chlorophpll a molecules had remained intact and that they had been linked into an amorphous ash by quadrivalent positive carbon atoms. In the event of such developments the amount of carbon necessary to effect such a linkage would be calculable as one quadrivalent sub-ionic atom per nucleus, or one carbon atom per molecule. It was obvious that such a development would account for the value C_{55} as interpreted from the data of conventional chemical analyses.

With the attainment of tentatively satisfying explanations of the differences between the contemporary chemical appraisal of chlorophyll a and the appraisal evolved in conjunction with the corollaries of the description of periodic hydrational potentiality the projected structural formula given in Figure 2 definitely attained stature in the direction of an improved configuration.

The outstanding feature of the molecular structure projected in Figure 2 was its symmetry; a feature made more obvious by a direct comparison with the molecular structure projected in Figure 1. Analogy was to be recognized as a sharp dangerous tool and yet there was the inevitable suggestion that many structural molecular formulas projected solely on the basis of potential chemical reactivity might involve impaired molecules and hence might represent pathological aspects.

Another important feature of the structure projected in Figure 2 was its high degree of correlation with observationol analytical data and the consequent validation of the precise change in weight with ionization which had been so repeatedly evidenced by the behavior

patterns of solutes and gases. To an appreciable extent the outstanding contributions of chemical science had seemed to preclude
the possibility of error in conjunction with the conventionat interpretations of data relating to atomic weight. Yet herewith the
analytical data of chemical science had appeared to reveal the
asymmetry of such a formula as that given in Figure 1 as attributable
in part to a simple failure to fully characterize the sub-ionic atomic
components of a molecule and in part to a failure to make appropriate allowances for change in weight with ionization. These considerations rather clearly seemed to justifiy further re-examinations of
data basic to conventional appraisals.

Empirically the modification of chlorophyll a which led to the
development of chlorophyll b was contemplated as a change which
insured the persistent presence of the catalytic constituents of a
water vapor molecule in the transition of plants from an aqueous
to an aerial environment. The change was ventured as having taken
place within the nucleus of the chlorophyll a molecule and as
having involved the addition of the component sub-ionic atoms of
an H_2O molecule to the nuclear field of major resonance. The change
has been indicated in Figure 3.

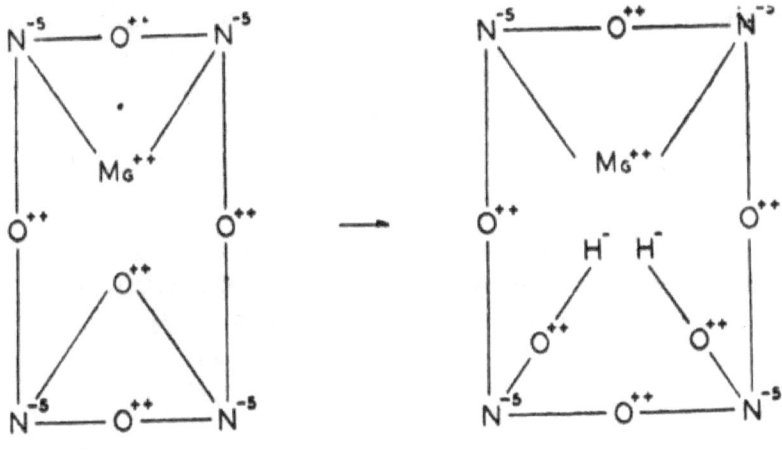

Figure 3.   Projected nuclei for chlorophylls a and b.

The modification projected in Figure 3 yielded for chlorophyll b a compositional formula $C_{32} H_{74} O_6 N_4 Mg$, a formula differing from that projected for chlorophyll a by the unit atoms in an H_2O molecule. Although the composition as projected still represented an unexplained departure from the conventional compositional formula for chlorophyll b it was considered more reasonable to venture that the modification of chlorophyll a had not involved any impairment of the peripheral complement of 72 hydrogen atoms.

Originally the inquiry herewith documented had been directed to a study of chlorophyll, but the symmetrically stellate structural pattern given in Figure 2 seemed to hold something of the infinite charm of snowflakes and prompted an extension of the inquiry to the associated pigments carotin and xanthophyll.

CAROTIN AND XANTHOPHYLL

It was to be noted that on the basis of conventional compositional or structural formulas there was no obvious relationship betweet chlorophylls a and b on the one hand and carotin and xanthophyll on the other hand. Yet analyses of green tissues had evidenced the rather uniform occurrence of the latter orange and yellow pigments and in consequence there was the suggestion that carotin and xanthophyll were normal stages in the synthesis of the chlorophylls in the higher plants. Since the blue-green algae were devoid of xanthophyll it was ventured that carotin was the primary pigment and was thus comparable with chlorophyll a in the chloro phyll group.

When a structural formula was projected empirically for carotin from the background used in the derivation of the structural formula for chlorophyll a it was found that insofar as carbon and hydrogen were concerned six of the eight unit aggregations indicated in Figure 2 as surrounding the nucleus of the chlorophyll a molecule yielded a composition commensurate with the data of chemical analyses. These six units yielded a compositional formula of $C_{24} H_{84}.$

Eighteen of the sub-ionic C^{-4} atoms in the absence of an allowance for change in weight with ionization would have been interpreted as thirty-six carbon atoms and the resulting formula thus would have become $C_{42} H_{54}$. It was obvious that these six units would have to be bonded in some way to form a carotin molecule. The most likely bonding agent seemed to be oxygen, since oxygen was recognized as having a dynamic role in the associated chlorophylls.

As projected, the unit aggregations of carbon and hydrogen comprised the common building blocks of all our protoplasmic pigments, on which account carotin and xanthophyll became subject to appraisal as stages in both the synthesis and the breakdown of the chlorophylls. These unit aggregations thus might be ventured as relatively stable radicals in nature. A structural formula for one of the units has been given in Figure 4.

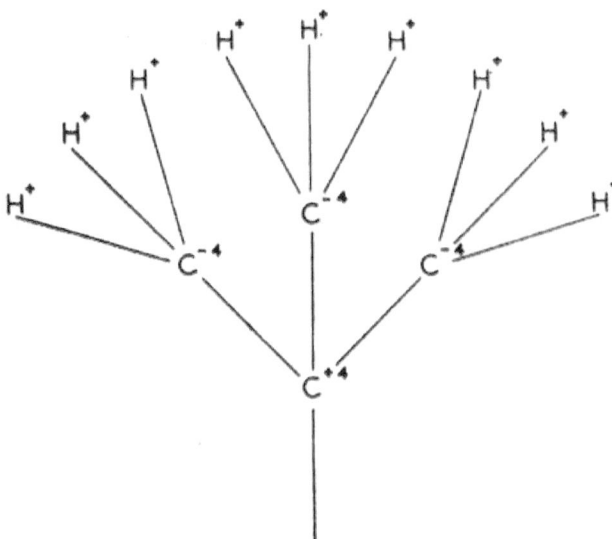

Figure 4. Structural formula for one of the eight composite sub-ionic units peripheral to the nucleus in the molecules projected for Chlorophylls *a* and *b*.

The projection of oxygen as the most likely bonding agent unit-
ing six of the units indicated in Figure 4 to form carotin required
that as a sub-ionic atom oxygen would have to have a complement
of six peripheral electrons. Yet the data for the chemical composition of carotin had indicated that 56 atoms of hydrogen were
presnt, — and the six units contained only 54 atoms of hydrogen. In
recognition of the analytical data it was ventured that oxygen as the
bonding agent in carotin possessed a full complement of eight peripheral electrons. As thus ventured the structural formula
for carotin became subject to projection as represented in
Figure 5.

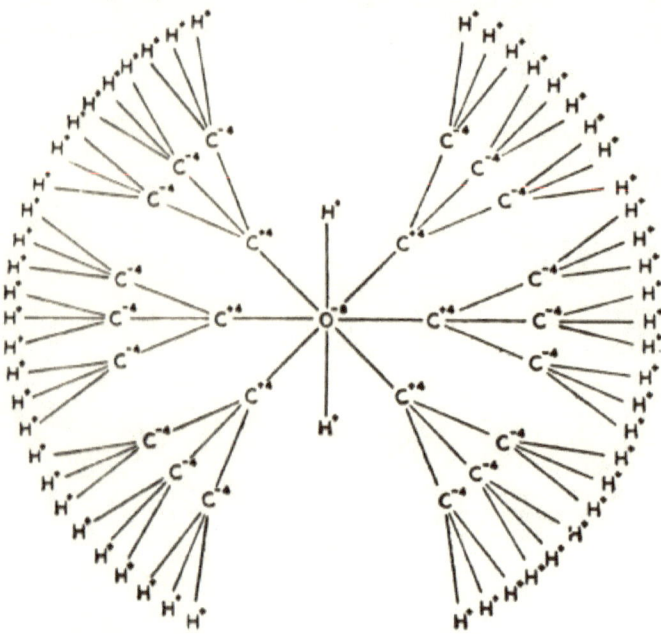

Figure 5. Projected structural formula for carotin.

In the foregoing consideration of the two chlorophyll pigments
chlorophyll *a* was appraised as the primary pigment in geologic time

and chlorophyll *b* was appraised as a modification attributable to changes in environment: the gradual clearing of the earth's hydrosphere, the consequent increase in the amount of solar radiation reaching the earth's surface, the correlated transition from an aqueous to an aerial habitat.

An analogous development was projected as having taken place with respect to carotin also appraised as a primary pigment in geologic time. As in the case of the chlorophylls the modification was projected as restricted to the central portion of the molecule. A diagram illustrating the nature of the projected modification of carotin has been given in Figure 6.

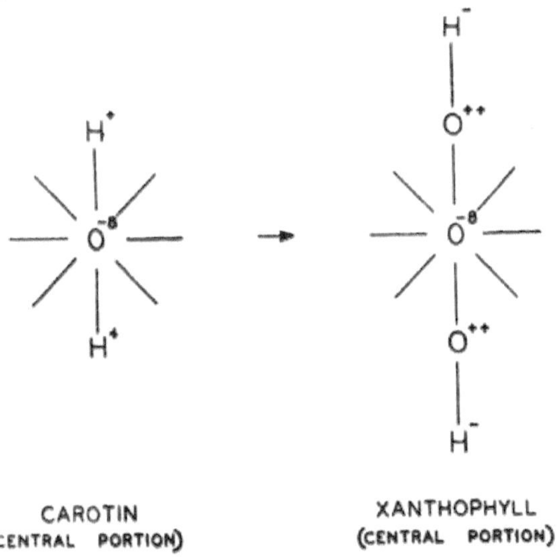

CAROTIN
(CENTRAL PORTION)

XANTHOPHYLL
(CENTRAL PORTION)

Figure 6. Projected transition of the carotin molecule (central portion only) to the xanthophyll molecule (central portion only).

The compositional formulas for carotin and xanthophyll as represented in Figure 5 and 6 were to be noted as not in strict conformity with contemporary interpretations in chemical science and yet there were ready explanations for the differences. The use of an 8—valent negative sub-ionic atom of oxygen as a bonding agent was unusual, but it was of interest to note that as a free ion such a unit would be *without weight* and hence as such would not be subject to gravimetric detection in analysis. As noted, a failure to recognize change in weight with ionization would yield 42 instead of 24 as the number of carbon atoms in the molecules of each pigment. The conventional value 40 suggested that incident to the analytical procedures the degradation of the molecule by heat two of the sub-ionic C^{-4} atoms had united with a sub-ionic O^{-8} atom to form a residual ash,—a development analogous to the suggested union of one sub-ionic C^{+4} atom with the nucleus in the case of chlorophyll *a* and chlorophyll *b*.

As represented in the foregoing considerations the four protoplasmic pigments chlorophyll *a*, chlorophyll *b*, carotin and xanthophyll seemed sufficiently correlated and integrated to justify the direction of attention to allied aspects relating to their occurrence in nature and to their patterns of behavior.

RESONANCE AND FLUORESCENCE

It was recognized that the projections given for the structural formulas of chlorophyll *a*, chlorophyll *b*, xanthophyll and carotin all included peripheral sub-ionic atoms of hydrogen. It was recognized also that sub-ionic atoms of hydrogen in water vapor in the earth's atmosphere had clearly been evidenced as capable of absorbing visible radiation at appropriate specific wave-lengths: the wave-lengths of certain Frauenhofer lines to be discussed at a later point. It was ventured, therefore, that in each of the four pigments absorption of radiation by peripheral hydrogen would account for the basic common features of their absorption of short-wave visible radiation.

In contrast to the basic common absorption it was ventured that absorption associated with resonance was mediated within the nucleus of the chlorophyll *a* and *b* molecules by sub-ionic atoms of oxygen, magnesium and hydrogen. Common to both chlorophyll *a* and chlorophyll *b* was a sub-ionic atom of magnesium appraised as within a field of force and highly subject to resonance by radiation absorbed at appropriate specific wave-lengths. In chlorophyll *a* a sub-ionic atom of oxygen similarly within a field of force was appraised as highly subject to resonance. In chlorophyll *b* two sub-ionic atoms of hydrogen similarly within a field of force were appraised as subject to resonance. For reasons to be considered at a later posnt the chemical activity of chlorophyll was interpreted as mediated by resonance, which in turn, was induced by radiation absorption at specific wave-length.

A unique feature of the chlorophylls was their fluorescence, inasmuch as the quality of their fluorescent radiation was such as to be maximally absorbed in the wave-length region in which absorption by the resonant atoms of hydrogen and oxygen was most active. In effect it thus was subject to projection that radiation absorbed by the chlorophyll pigments at wave-lengths shorter than the red region of absorption yielded fluorescent radiation of a quality to be most effective at the nuclear site of the potentially resonant sub-ionic atoms of oxygen and hydrogen. Allied with this projection was the contemplation of sub-ionic magnesium whose resonance was incited by specific radiations poorly absorbed by any other sub-ionic constituents of chlorophyll.

SOME CRITICAL WAVE-LENGTHS OF ABSORPTION

In the earth's atmosphere the absorption of solar radiation by gaseous molecules has been documented by the positions of shadows within a spectrum. These shadows have been termed Frauenhofer lines and the major lines have been given alphabetical designations. It was of interest to note that sub-ionic atoms in molecules absorbed radiation in a manner characteristic of their specific elemental nature. The Frauenhofer lines involving absorption by gaseous oxygen, hydrogen and magnesium have been assembled to comprise Table 1.

Table 1. The Frauenhofer lines of absorption by gaseous oxygen, hydrogen
and magnesium in the earth's atmosphere.

	Designation of Frauenhofer Line	Wave-length in Angstrom Units	Absorbing Element
Section I.	A	7593.8	Oxygen
	B	6867.2	Oxygen
	C	6562.8	Hydrogen
	F	4861.4	Hydrogen
	G	4340.5	Hydrogen
	H	4101.9	Hydrogen
Section II.	b_4	5167.3	Magnesium
	b_2	5172.7	Magnesium
	b_1	5183.6	Magnesium

It was to be recognized that in the absorption of radiation by
solids and liquids the specific nature of the absorbing material was
obscured by factors associated with the large number of atoms
present. The absorption of spectral radiation by leaves had yielded
shadows in broad bands rather than in lines, especially in the violet-
blue region, and it was largely on this account that both pigment
absorption and the correlated photosynthetic activity commonly had
come to be considered as having involved solids or liquids. The
Grotthus law had held that only absorbed radiation was effective in
the excitation of a response. Historically the law played a substantial role in the analysis of plant responses, but in recent years the
recognition of fluorescence has done much to nullify its practical
significance and application.

As noted previously the four protoplasmic pigments were insoluble in water but were soluble in various organic solvents. In
solution it was obvious that as the concentration was decreased the
solutes became more and more analogous to gases in their attendant
degree of freedom. Yet when the same pigments were dissolved in
different solvents it was found that the critical wave-length of their

absorption *varied with the solvent used*. This meant that no very precise significance safely could be attached to the specific observed critical peaks of the absorption data. With this information in mind it became of interest to compare the data on Table 1, Section 1, with the absorption peaks obtained for ether extracts of the chlorophylls by Zscheile and Comar. The data have been made available for comparison in Figure 7.

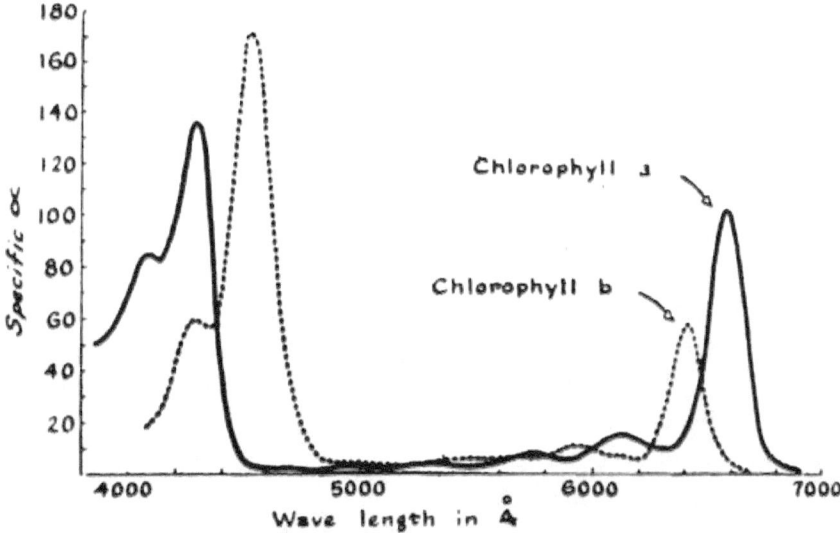

Figure 7. Data indicating the absorption peaks of chlorophyll *a* and *b* in either solution.

When the absorption peaks given in Figure 7 were compared with the absorption data given in Table 1 it became obvious that a significant correlation was possible, especially since the peaks given in Figure 7 were subject to variation with different solvents. In Figure 7 the degree of steepness of the curves approaching the peaks from either side was to be interpreted as dependent primarily upon the concentration of the solute. As the concentration was reduced the relative freedom of the solute units was increased and their absorption progressed in the direction of the linear absorption characteristic of gases. From the indicated degree of potential correlation it seemed reasonable to conclude that in chlorophyll the absorption of visible radiation was largely attributable to oxygen and hydrogen. It was to be noted that this conclusion was entirely consistent with the resonance indicated for oxygen and hydrogen in the projected symmetrical structural formulas. It was to be noted that no absorption peak attributable to magnesium was evidenced,— a situation which suggested that resonant magnesium was far more stable than the associated oxygen and hydrogen, and that its role was strictly catalytic.

SOME CRITICAL WAVE-LENGTHS OF PLANT RESPONSE

With the Grotthus law in mind it was natural that attempts were made to establish direct correlations between the absorption characteristics of specific contained pigments and specific reactions to the absorbed radiation. For example, Hoover studied the action spectrum of wheat leaf photosynthesis but obtained results which indicated that photosynthesis took place on exposures to radiation very poorly absorbed by chlorophyll. Such results became understandable following evidence of the extraordinary sensitivity of chlorophyll to the fluorescent re-radiation of chlorophyll. As noted previously, the application of the Grotthus law was unfortunate whenever fluorescence was involved. In the case of chlorophyll an exposure to blue light would incite fluorescent red light which was maximally effective in photosynthesis.

With extraordinary prejudice it was ventured that to date the most satisfying data on the critical wave-lengths of plant response had been obtained with small seeds, such as lettuce. The most important aspects of these studies were subject to appraisal as (1) the establishment of a critical wave-length near 6800A as the radiation most effective in promoting the germination of the seeds (2) the establishment of a critical wave-length near 7600A as the radiation most effective in inhibiting the germination of the seeds and (3) the repeatable reversibility of the seed reactions to the indicated radiations. The importance of these developments was their relation to the data given in Figure 7 and Table 1. There was no question but that chlorophyll a had been involved in the seed reactions to light, since chlorophyll had been found to be present in the seeds and the two critical wave-lengths regions were those of the A and B Frauenhofer lines of absorption by oxygen. The reversibility of the seed reaction potentialities also was consistent with the involvement of a resonant sub-ionic atom of oxygen, thereby serving to enhance the plausibility of the projected structural formula for chlorophyll a.

It was obvious that, as interpreted, the absorption of radiation at 7593.8A by oxygen in the nucleus of the chlorophyll a molecule incited an inhibition directly opposite to the promotion incited by the absorption of radiation at 6867.2A by the same oxygen. On this account it became of interest to conduct a study of the effect of the development of seedlings when radiation at the 7600A region was superimposed on radiation at the 6800A region. It was found in such studies that the height growth was restricted and that the leaves became a darker green in the area to which 7600A radiation had been added. There was the suggestion that the 7600A radiation in nature might function as a governing or restraining force.

GEOLOGICAL ASPECTS

The blue-green algae of today, appraised as persistent souvenirs of a very remote geological era, were of special interest in a study of protoplasmic pigments because they contained chlorophyll a and carotin, but were devoid of chlorophyll b and xanthophyll. The era of their dominance on the earth has been variously viewed by geologists but the general consensus of opinion has been that it was an era of relatively low light intensity during which solar radiation was intercepted by one or more hydrospheric envelops. At the present time the blue-green algae have been observed as thriving at light intensities appreciably below those in the open on cloudless days. There was the suggestion, therefore, that chlorophyll *a* might be extremely sensitive to the red light which in a gradually clearing hydrospheric habitat would have been ths first visible light to reach the earth's surface from the sun.

The suggestion prompted the carrying out of an experiment designed to test the order of sensitivity of chlorophyll to red light. The results obtained in this experiment, carried out at the Boyce Thompson Institute for Plant Research, indicated that fluorescent red light of an extremely low intensity, induced by spectral ultra-violet radiation impinging on a heavy plate of glass, could initiate the germination of light-sensitive lettuce seeds. The data have been given in Figure 8.

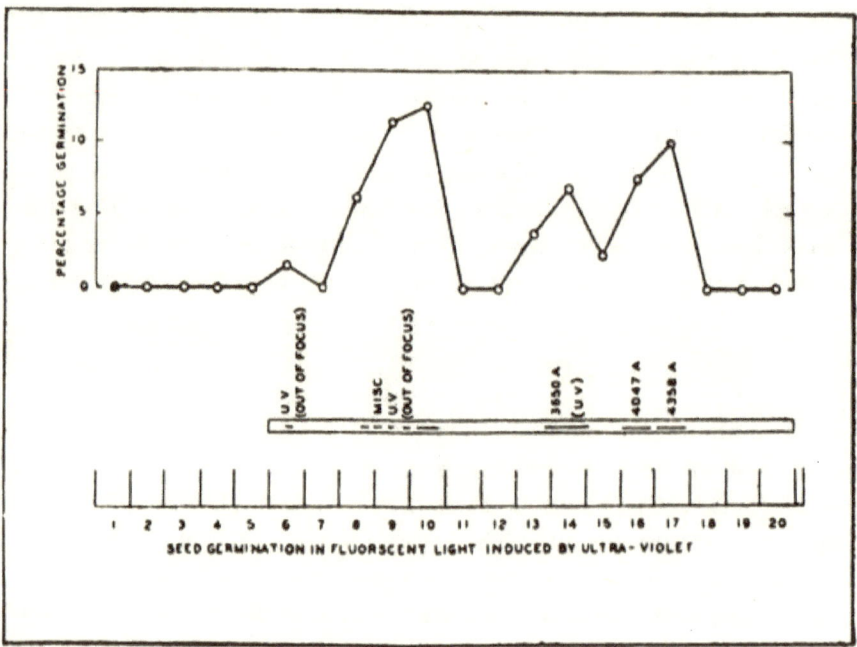

Figure 8. Data obtained in a study of the effect of low-intensity red light, induced as fluorescence, on the germination of light-sensitive lettuce seed.

In Figure 8 the ultra-violet radiation originated from a carbon arc source approximately twelve feet away. The radiation was passed through quartz collimating lenses and a quartz prism, following which the invisible spectral radiation was reflected twice from special surface-coated mirrors. It was then allowed to impinge upon a heavy Corning glass fluorescing red. The energy of the fluorescent radiation emergent below the glass was estimated as being about one twentieth of the imposed radiation. Under the conditions of the experiment the energy of the radiation varied with the particular wave-length involved but was appraised as low in all cases even before impingement on the fluorescent glass. Yet germination was incited,—and previously it had been made evident that chlorophyll had been involved in the germination of the seeds.

The results given in Figure 8 seemed very suggestive with respect to the appraisal of chlorophyll *b* as a modification of chlorophyll *a* facilitating the effectiveness of the shorter wave-length visible radiation postulated as attending the modification of the earth's projected hydrospheric envelop. Indirectly also, the results were suggestive with respect to the concept that xanthophyll was a modification of carotin which took place incident to the gradual clearing of the earth's hydrosphere and the attendant transition of plant life from an aqueous to an aerial environment. As represented in the projected structural formulas the modifications of both chlorophyll *a* to chlorophyll *b* and carotin to xanthophyll insured the presence of the constituents of an H_2O unit as a catalyst in the absence of an aqueous environment.

PATHOLOGICAL ASPECTS

Inasmuch as xanthophyll and carotin as well as the chlorophylls were insoluble in water it was to be recognized that under laboratory conditions the solutions of xanthophyll and carotin might be degenerative desivatives of chlorophyll formed by the actions of the organic solvents. In general the relative amounts of the yellow and orange pigments, as compared with the chlorophylls, increased with the age of the plant tissue,—a relationship which seemed to carry a suggestion that a deterioration of chlorophyll was involved. In healthy green leaves it was to be noted that xanthophyll and carotin, if present, were obscured by chlorophyll. In non-healthy leaves the presence of yellow and orange pigments was as attributable to xanthophyll formed in the denaturalization of chlorophyll as to the simple loss of chlorophyll projected as having masked xanthophyll and carotin already present.

The natural presence of obvious yellow and orange pigments in plants without exception was found to be subject to interpretation as associated with degenerative agencies. The excellent researches of Lutman yielded results which emphasized the viewpoint that in numerous instances the development of fleshy tubers and roots was intiated by systemic mildly pathological microorganisms. Two of the common plant sources of carotin,—carrots and sweet potatoes— were subject to inclusion within this category. The development of yellow and orange pigments in conjunction with virus diseases of plants was subject to interpretation as having involved a breakdown of chlorophyll by one or more virus toxins. The occurrence of such pigments in variegated plants similarly was attributable to the presence of one or more mild systemic viruses. There was evidence that dodder plants, commonly with abundant carotin, contained one or more systemic viruses. Even in the ripening of oranges the increase in carotin has been evidenced as correlated with a decrease of chlorophyll. The incidence of xanthophyll and carotin in certain fungi and bacteria readily was subject to interpretation as having involved pathological agencies and degenerative processes.

DISCUSSION

In retrospect it seemed appropriate that historically chemical research had directed attention primarily to the composition of matter: its atoms and molecules. Such research led naturally to studies of the interrelationships of these units of matter. To what might seem an unfortunate extent the study of the relationship of these units of matter to an aqueous solvent was neglected.

The discovery of the periodic nature of hydrational potentiality, as was made evident in a recent publication, laid the ground work for a new approach to various aspects of solute behavior. In general, hydration afforded a key to a wider recognition of symmetry in solute associations. The key was worthless without the analyses of chemists, but with these analyses as a guide there was at hand a new pathway to progress in areas long appraised as formidable.

With respect to chlorophyll the discoveries relating to hydration indirectly contributed not only to the projection of such symmetrical molecules as the one indicated in Figure 2 but also supplied the background for suggestions relating to the behavior patterns of molecules. As regards photosynthesis, osmosis had evidenced a potentiality for effecting the breakdown of solute carbonate radicals and hence had been indicated as a mechanism facilitating carbohydrate synthesis in the chlorophyll nucleus. The osmotic behavior of solute nitrate and ammonium radicals when accompanied by copper sulfate, as reported in the mentioned publication, clearly evidenced their breakdown into atomic ions. It was to be recognized that in chlorophyll the magnesium ion Mg^{++} was subject to replacement by the copper ion Cu^{++}, on which account there was at hand in chlorophyll a mechanism facilitating protein synthesis. It was of interest that magnesium sulfate failed to evidence any breakdown of the nitrogen-containing radicals, nor was copper nitrate effective.

As illustrated by the studies herewith reported on chlorophyll and its related protoplasmic pigments, the most promising contribution of the description of periodic hrdrational potentiality was its significance with respect to categories of weight, to change in weight with ionization and to the status of sub-ionie atoms in molecules.

The involved principles constitute new pathways to greater understanding, – pathways which envision for organic chemistry the recognition of a new order of symmetry in the projected interpretations of complex molecules.

SUMMARY

The present appraisal of chlorophyll evolved from the application of two principles inherent in a description of periodic hydrational potentiality. These principles were (1) the full characterization of the sub-ionic atoms in a molecule and (2) a change in weight with ionization. The two principles were not in conformity with contemporary chemical science, but repeatedly had been sustained as valid by the behavior patterns of solute ions. The structural formulas projected through adherence to these principles were uniformly symmetrical, in sharp contrast to conventional projections derived by inference from chemical reactions and hence subject to interpretation as having involved impairments of the molecules.

Chlorophyll *a* was appraised as the basic photosynthetic pigment and chlorophyll *b* as a modification chronologically attendant upon the transition of plants from an aqueous to an aerial environment and from an early era of reduced light intensity to an era comparable with present conditions.

Carotin was appraised as a pigment representing a stage in the synthesis and degradation of chlorophyll. Xanthophyll was appraised as a modification of carotin associated with the same environmental changes which had been involved in the modification of chlorophyll *a*.

Plant reactions involving chlorophyll were appraised as attributable to the absorption of radiant energy by resonant oxygen and hydrogen in the nuclei of chlorophylls *a* and *b*, coupled with the catalytic activity of associated resonant magnesium.

The extreme sensitivity of chlorophyll to radiation at a wavelength of 6867.2 Angstroms was appraised as evidence that fluorescent re-radiation from chlorophyll exposed to shorter wave-length radiation took part in the induction of plant responses.

A reversible plant reaction to radiations in the longer wavelength spectral region was interpreted as attributable to absorption by resonant oxygen in the nuclei of chlorophyll a molecules at wave-lengths of 6867.2A° and 7593.8A° respectively.

The parallel absorption characteristics of the four protoplasmic pigments in the shorter wave-length spectral region were interpreted as attributable to absorption by peripheral sub-ionic hydrogen atoms common to each pigment.

The natural exclusive or predominant occurrence of carotin, xanthophyll or both was correlated with agencies potentially capable of effecting a denaturalization of chlorophyll molecules or of interfering with the synthesis of chlorophyll molecules.

REFERENCES

Bonner, James, and Arthur W. Galston (1952).   Principles of Plant Physiology. Freeman and Co.

Dufrenoy, Jean (1942).   Researches in the Department of Botany, Bacteriology and Plant Pathology. Louisiana State University, Baton Rouge, La.

Fischer, Hans, and A. Stern (1940). Die Chemie des Pyrrols. II. Akademische Verlagsgesellschaft. [Leipzig].

Flint, Lewis H. (1963).   Behavior Patterns of Hydration. Inst. for the Advancement of Science and Culture, New Delhi, India.

— (1952).   Note on Variegation and a Virus Disease of Dodder.   Proc. La. Acad. Sci. 15:17-19.

— and C. F. Moreland (1939).   Responses of Lettuce Seedlings to 7600A Radiation.   Amer. Jour. Bot. 26:231-233.

— (1938).   Immediate Problems in the Light-sensitivity of Plants. Spectroscopy in Science and Industry. Wiley & Sons.

—, and E. W. McAlister (1937).   Wave Lengths in the Visible Spectrum Promoting the Germination of Light-sensitive Lettuce Seed. Smithsonian Miscellaneous Collections 96 (No. 2).

— and E. D. McAlister (1934).   Wave-Lengths of Radiation in the Visible Spectrum inhibiting the germination of Light-Sensitive Lettuce Seed. Smithsonian Miscellaneous Collections 94 (No. 5).

— (1933). Hydration of the solute ions of the heavier elements. Journ. Wash. Acad. Sci. 22:211-217.

— (1932). Hydration of the solute ions of the lighter elements. Jour. Wash. Acad. Sci. 22:97-119.

Hoover, W. W. 1937). The Dependence of Carbon Dioxide Assimilation in a Higher Plant on Wave Length of Radiation. Smithsonian Miscellaneous Collections. 95 (No. 21).

Lutman, B.F. (1945). Actinomycetes in Various Parts of the Potato and Other Plants. Vt. Agri Exp. Sta. Bull. 522.

Willstatter, Richard (1910). Chlorophyll und seine wichtigsten Abbauprodukte. Abderhalden's Handb. biochem. Arbeitsmeth. 2: Berlin.

— and Ernest Hug (1911). Isolierung dee Chlorophylls. Leibig's Ann. Chem. u. Pharm. 380:177-211.

— and Arthur Stoll (1933). Untersuchungen über Chlorophyll, Methoden und Frgebnisse. Berlin.

Zscheile, F. P., and C. L. Comar (1941). Influence of Preparative Procedure on the Purity of Chlorophyll Components as Shown by Absorption Spectra. Bot. Gaz. 102:463-481.

CHAPTER 9.   CONCERNING COLOR AND HYDRATION

The present chapter describes some experiments in which color appeared intimately related to hydrational potentialities. Because the experiments were of a very simple nature it was obvious that they afforded to any interested person a ready means of personal testing. Because of the far-reaching and dramatic significance of the description of periodic hydrational potentiality with its four corollaries characterizing categories of weight any procedure contributing to a wider recognition of the integrity of the description was considered of value in helping to extend the frontiers of science.

SOME BEHAVIOR PATTERNS OF COBALT CHLORIDE

For many years cobalt chloride has been recognized as an indicator of the presence or absence of water vapor. In aqueous solution a deep pink color was exhibited, and since aqueous solutions of many chlorides were colorless it was natural to associate the color with the presence of cobalt ions. Aqueous solutions of $CoCl_3$ and $CoCl_2$ had the same pink color. Substances absorbing these solutions and thereby attaining a pink color became blue upon drying out. To some extent such materials have been used as "weather indicators," registering by color changes from pink to blue or blue to pink some changes in atmospheric moisture. In plant physiology papers soaked in weak solutions of cobalt chloride and dried out to a blue color have been used in the study of transpiration from leaves. More recently aqueous solutions of cobalt chloride have been used in studies of the amount of so-called bound water in plants.

In the experiments here reported the hydrated salt $CoCl_2.6H_2O$ was used. Both the salt and its aqueous solutions were of a pink color. Papers dipped in solutions were pinkish, and became bluish upon being dried out.

Under the circumstances it was natural to conclude that the cobalt ion Co^{++} was of such a nature as to appear pinkish when hydrated and bluish when anhydrous. In line with such a conclusion it was of interest to use the description of periodic hydrational potentiality and calculate respective values for the assumed ions involved. In this procedure cobalt with the atomic number 27 had a calculated weight of 54 in the neutral state and a calculated weight of 58 as the ion Co^{++}. As an ion of weight 58, cobalt would have, from the description, an ability to take on 17 H_2O-units in hydration. By analogous procedures the Cl^-ions would have a weight of 32 and an ability to take on 7 H_2O-units in hydration. As a solution, therefore, the average hydrational potentiality was subject to calculation as follows:

$$Co^{++} = 17 \ H_2O^- \quad Cl^- = 7 \ H_2O^- \quad Cl^- = 7 \ H_4O$$
$$\frac{17 + 7 + 7}{3} = \frac{31}{3} = 10.33$$

The value thus derived, 10.33, was appraised as representing, on a relative basis subject to comparison with values for other solutes similarly derived, the ability of the solute ions of cobalt chloride, $CoCl_2$, to retain H_2O-units. Previously published data relating to the specific gravity of aqueous solutions had indicated very definitely that this retentive ability or hydrational potentiality persisted in the absence of free solvent and under such conditions became a force effecting bondage. An entire chapter in the book "Behavior Patterns of Hydration" was devoted to a consideration of hydrational bondage.

It became of interest at this time to utilize the description of periodic hydrational potentiality in a similar manner to calculate values for other solutes which might then be compared with the value 10.33 which had been derived for cobalt chloride, $CoCl_2$. In this project the solutes selected were the chlorides of potassium, sodium and lithium. It was assumed that in aqueous solution the ions involved would be Li+, Na+, Cl-, and K+. For these ions the ionic weight values and the numbers of H_2O-units potentially held were subject to derivation from the description of periodic hydrational potentiality as follows, arranged in the order of increasing ionic weight.

Li+, ionic weight 8, number of H_2O-units 19

Na+, ionic weight 24, number of H_2O-units 11

Cl-, ionic weight 32, number of H_2O-units 7

K+, ionic weight 40, number of H_2O-units 3

From the values thus derived the average ability of the ions involved in the respective chlorides were calculated as follows.

For LiCl, $\dfrac{19+7}{2} = \dfrac{26}{2} = 13$

For NaCl, $\dfrac{11+7}{2} = \dfrac{18}{2} = 9$

For KCl, $\dfrac{3+7}{2} = \dfrac{10}{2} = 5$

For convenience in comparison the respective calculated values for these three salts and for cobalt chloride were arranged in the order of decreasing average hydrational potentialities.

LiCl —Li+, Cl-   —13

$CoCl_2$ —Co++, Cl-, Cl-   —10.33

NaCl —Na+, Cl-   —9

KCl —K+, Cl-   —5

Upon comparing the values derived by means of the description of periodic hydrational potentiality for aqueous solutions of KCl, NaCl and LiCl with the value similarly derived from $CoCl_2$ it was obvious that whereas $CoCl_2$ had a greater retentive value than KCl or NaCl, the value for LiCl exceeded that of $CoCl_2$. To aliquot pink solutions of $CoCl_2$ anhydrous salts of the three chlorides were added. The results obtained were as follows:

KCl + $CoCl_2$ solution: remained pink

NaCl+CoCl$_2$ solution: remained pink
LiCl+CoCl$_2$ solution: turned blue

From these results it was concluded that the lithium chloride had brought about the dehydration of the previously hydrated cobalt ion, that potassium and sodium chlorides had been unable to bring about such a dehydration, and that the results were consistent with the values calculated from the description of periodic hydrational potentiality.

If the interpretation was valid it was of special interest for correlation with previously published data. The blue color of the liquid present following the addition of appreciable amounts of lithium chloride to the pink solution of cobalt chloride carried the obvious implication that no free solvent was present. This was precisely the implication which was inherent in the successful calculation or prediction of many specific gravity values for so-called aqueous solutions at high solute concentrations. In general and heretofore, however, there had not been available any tangible index of the absence of free solvent beyond the implication. Superficially the distinction between a true aqueous solution and one in which no free solvent was present and the solute ions were bonded by their respective hydrational potentialities for the H$_2$O⁻units present, though not free, was not obvious nor recognized. Previous to the description of periodic hydrational potentiality it was not even subject to calculation from the data of specific gravity. Regrettably the failure to recognize the transfer of solvent to solute was largely responsible for the interpretation of electrical conductivity data which led to the theory of incomplete dissociation.

In a practical way a simplification of the foregoing tests may be carried out by sprinkling a few crystals of LiCl, NaCl and KCl on a paper reddened by a solution of cobalt chloride. A blue color will appear at the points of contact of the lithium chloride, but not elsewhere.

SOME BEHAVIOR PATTERNS OF COPPER SALTS

The results which had been obtained in the studies reported prompted an extension of the inquiry. In this project two salts of copper were used : copper sulfate and copper nitrate. Each of these salts had received some attention in conjunction with previous investigations. The specific gravity of aqueous solutions of copper sulfate had been such as to evidence a complete dissociation of any projected sulfate radical. The osmotic behavior of copper sulfate had been such as to evidence a complete dissociation of any projected sulfate radical. The specific gravity and osmotic behavior of copper nitrate in aqueous solution had been such as to evidence the presence of intact nitrate radicals. Because of the results obtained in these earlier studies it was considered advisable to calculate average ionic hydrational potentiality values for copper sulfate on each of two assumptions: (1) that the aqueous solution contained the ions Cu^{++} and SO_4^{--}, and (2) that the aqueous solution contained the ions Cu^{++}, S^{+6}, O^{--}, O^{--}, O^{--} and O^{--}. With respect to copper nitrate it was assumed that aqueous solutions contained the ions Cu^{++}, NO_3^- and $NO_3.^-$

Using the description of periodic hydrational potentiality, the following procedures yielded the desired average values.

For $CuSO_4$ as Cu^{++} and SO_4^{--} in aqueous solution.

$$Cu^{++} = 62 + 15 H_2O^- \quad SO_4^{--} = 92 + 23 H_2O^-$$

$$\frac{15+23}{2} = \frac{38}{2} = 19$$

For $CuSO_4$ as Cu^{++}, S^{+6}, O^{--}, O^{--}, O^{--} and O^{--} in aqueous solution

$$Cu^{++} = 62 + 15 H_2O^-; \quad S^{+6} = 44 + 1 H_2O^-; \quad O^- = 12 + 17 H_2O^-$$

$$\frac{15+1+17+17+17+17}{6} = \frac{84}{6} = 14$$

For $Cu(NO_3)_2$ as Cu^{++}, NO_3^- and NO_3 in aqueous solution.

$$Cu^{++} = 62 + 15 H_2O^- \quad NO_3^- = 60 + 16 H_2O^-$$

$$\frac{15+16+16}{3} = \frac{47}{3} = 15.66$$

To facilitate comparison with $CoCl_2$ the involved average values were arranged in decreasing order.

$CuSO_4$ as Cu^{++} and SO_4^{--} — — — — — — — — — — 19

$Cu(NO_3)_2$ as Cu^{++}, NO_3^- and NO_3^- — — — — — —15.66

$CuSO_4$ as Cu^{++}, S^{+6}, O^{--}, O^{--}, O^{--} and O^{--}——14

$CoCl_2$ as Co^{++}, Cl^-, Cl^- — — — — — — — — — — — —10.33

When the respective copper salts were added to aliquots of $CoCl_2$ in aqueous solution it was found that both copper sulfate and copper nitrate changed the color from pink to blue:

$CuSO_4 + CoCl_2$ solution: pink to blue

$Cu(NO_3)_2 + CoCl_2$ solution: pink to blue

The results obtained were considered to be consistent with the calculated average values for hydrational potentiality and with the results which had been obtained with the chlorides of lithium, sodium and potassium. The above results, however, did not permit any diagnosis of the ions present in aqueous solutions of copper sulfate,—a situation which prompted a supplementary study.

SOME BEHAVIOR PATTERNS INVOLVING MANGANESE

In the course of a search for some solute which would yield colored ions of such a nature as to indicate the ions present in an aqueous solution of copper sulfate attention was directed to potassium permanganate, a salt which yielded a brilliant magenta color in aqueous solution. Such solutions had not been included in earlier studies and nothing was known as to the nature of the ions present. On this account two assumptions were made, one that an anionic radical MnO_4^- remained intact, the other that the radical became completely dissociated. The average hydrational potentiality values for the two assumed states were calculated through the use of the description of periodic hydrational potentiality.

For $KMnO_4$ as K^+ and MnO_4^-

$K^+ = 40 + 3H_2O^- MnO_4^- = 112 + 13 \ H_2O^-$

$$\frac{3+13}{2}=\frac{16}{2}=8$$

For $KMnO_4$ as K^+, Mn^{+7}, O^{--}, O^{--}, O^-, O^{--}

$K^+=40+3\ H_2O^-$ $Mn^{+7}=64+14\ H_2O^-$ $O^{--}=12+17\ H_2O^-$

$$\frac{3+14+17+17+17+17}{6}=\frac{85}{6}=14.16$$

To facilitate comparison of the above values with those similarly derived for copper sulfate and copper nitrate in aqueous solution the involved values were arranged in a decreasing order.

$CuSO_4$ as Cu^{++} and SO^{--} $=19$

$Cu(NO_3)_2$ as Cu^{++}, NO_3^-, NO_3^-, $=15.66$

$KMnO_4$ as K^+, Mn^{+7}, O^{--}, O^{--}, O^{--}, $O^{--}=14.16$

$CuSO_4$ as Cu^{++}, S^{+6}. O^{--}, O^{--}, O^{--}, O^{--} $=14$

$KMnO_4$ as K^+ and MnO_4^- $=8$

When the respective copper salts were added to aliquot aqueous solutions of potassium permanganate the results obtained were as follows.

$CuSO_4+KMnO_4$ solution: remained magenta

$Cu(NO_3)_2+KMnO_4$ solution: turned blue.

It was assumed that in the case of $Cu(NO_3)_2$ added to the $KMnO_4$ solution the dehydration of the Mn^{+7} ion had resulted in a loss of color an l that the resulting blue color was attributable to the Cu^{++} ion.

The results obtained were interpreted as indicating that in aqueous solutions of copper sulfate there were no intact sulfate radicals present. This was precisely in accord with the evidence afforded by numerous mea-urements of specific gravity. It was to be recognized, however. that on the same basis the specific gravity data for numerous other sulfates in aqueous solution had attested the presence of sulfate radicals. The osmotic behavior of copper sulfate also had indicated a complete dissociation of any assumed sulfate radicals, but this fact was without significance here, since osmosis commonly had been evidenced as effecting the complete dissociation of all sulfate radicals.

The results obtained also were interpreted as indicating that in aqueous solutions of potassium permanganate no intact permanganate radicals were present, since copper sulfate was unable to decolorize the solution. The fact that $Cu(NO_3)_2$ was able to dehydrate $KMnO_4$ and $CuSO_4$ was unable to do so seemed to supply further evidence that both $CuSO_4$ and $KMnO_4$ were completely dissociated in aqueous solution. As thus interpreted the presence or absence of potentialities for decoloration not only held significance with respect to hydration but also extended a basis for the diagnosis of the nature of solute ions present. With respect to $CuSO_4$ in aqueous solution it was of interest that the indicated state of complete dissociation was in precise agreement with the evidence supplied by the data of specific gravity.

DISCUSSION

The description of periodic hydrational potentiality was derived in 1932 as a mathematical extension of an inverse relationship between weight and hydration which had been reported in 1899 by ABEGG and BODLÄNDER. A detailed account is given in "Behavior Patterns of Hydration." In relation to the data given herewith it was to be recognized that the solute ions Li^+, Na^+, Cl^- and K^+ were included in the first of four periods of 23 units each. In each instance one-half the ionic weight plus the number of H_2O^- units potentially held in hydration equaled 23. The four-period system extended over the range 0 to 184 in terms of $O=16$, beginning with hydrogen and ending with uranium, and was thus integrated with the naturally-occurring elements. The ions of the first period were colorless, as were the salts derived from these ions incident to dehydration. The detection and measurement of hydration in such solute ions involved such attributes as electrical conductivity, specific gravity and osmotically-evidenced increases in volume.

In contrast to the situation prevailing in the first period quite a number of elements in the second period were colored. Among these the element cobalt was unique in that it was pinkish or reddish in the presence of moisture and bluish in the absence of moisture. Color thus became subject to appraisal as an index of status with respect to the presence of moisture or with respect to hydration.

In the absence of an accurate description of periodic hydrational potentiality it was considered doubtful that any satisfactory appraisal of the hydrational potentiality of the cobalt ions could have been made, even with color as an index, since the pinkish or reddish color persisted in concentrated solutions under which conditions the H_2O^- units were shared and thus served in a bonding capacity.

With the description of periodic hydrational potentiality, on the other hand, it was a relatively simple matter to calculate values for any ion of specified weight, and it had been demonstrated repeatedly that the hydrational potentiality was strictly a function of ionic weight, equally operative with atomic ions and radicals. These developments made possible the use of color as an index of hydration in the simple experiments described.

It was of interest that the ultimate effectiveness of any association of ions was subject to significant correlation with summations or averages representing the combined activity of all ions present, and not with the activity of any specific ions of seemingly predominant potentialities. The rate of reaction, on the other hand, appeared to be subject to correlation with relative individual potentialities of specific ions.

SUMMARY

Studies involving color changes and specific ions yielded results which were subject to intimate correlation with calculated values for hydrational potentialities. The results did not appear to be subject to interpretation on any alternative basis and thus were appraised as constituting supplementary evidence of the integrity of the involved description of periodic hydrational potentiality. The observed patterns of behavior when thus interpreted also proved indicative of the nature of the ions present in the involved aqueous solutions.

LITERATURE CITED

1. Flint, Lewis H. (1963a). Concerning Chlorophyll. Advancing Frontiers of Plant Sciences, Vol. 4. Institute for the Advancement of Science and Culture, New Delhi, India.
2. — (1963b). Behavior Patterns of Hydration. Advancing Frontiers of Plant Sciences. Institute for the Advancement of Science and Culture, New Delhi, India.

CHAPTER 10.   CONCERNING TURGOR

The book entitled "Behavior Patterns of Hydration" made available for the first time an adequate mathematical background for biological research in areas involving water relationships. Its basic contribution was a description of periodic hydrational potentiality which characterized the attraction between ions and H_2O-units over the weight range of the naturally-occurring elements. The four corollaries of the description defined categories of weight. For some phases of researches involving considerations of volume the usefulness of the description became appreciably enhanced by the discovery of the relation of hydration to density. Such a development, for example, was a prerequisite to successful interpretation and prediction with respect to the specific gravity of aqueous solutions and with respect to osmosis. The results obtained in researches in these areas in turn became suggestive and helpful in relation to other problems, including intimate problems in the life sciences. It became obvious that osmosis brought about the hydration of solute ions in such a manner as to increase solution volume and thus to create hydrostatic pressure. The discovery that osmosis had potentialities for the dissociation of many solute radicals and ionic non-electrolytes and that the units thus released also became hydrated seemed to be of special significance to biology, not only because it represented a mechanism for the release of elements suitable for metabolic syntheses but also because it engineered further and appreciable volumetric increases and the consequent development of hydrostatic pressures. The present paper reports on the results obtained in a survey of the mentioned principles and discoveries in their potential relation to the origin and maintenance of turgor.

INDIGENOUS ELEMENT IONS

The principles involved in the volumetric increases attendan upon the hydration of element ions were assumed to be equally applicable in the physiology of plants and animals. In plants the intake of solutes commonly has been considered more a matter of circumstance than of choice, notwithstanding the evidence for selecti- vity. In animals a variable amount of choice has appeared to be commonly associated with circumstance. Collectively plants and animals share no group of element ions as a premise to metabolism and yet the aqueous solvent of their intake may include such ions. A few such ions may be selected as representative, even though in any specific instance they might or might not be present.

For this survey the following eight ions were selected :
Ca^{++}, K^+, Cl^-, Mg^{++}, Na^+, Fe^{+++}, Cu^{++}, Zn^{++} .The potential contri- bution of hydration to increase in volume was calculated for each of these ions and the results obtained were assembled to compromise Table 2.

From the data of Table 2 it was apparent that the hydration of element ions brought about appreciable increases in their volume. Since osmosis mediated the hydration of all of the ions listed, as had been evidenced by previously published data, it followed that such increases when taking place within the vacuoles of cells would produce hydrostatic pressure.

It was of interest that the sodium ion, Na^+, possessed the greatest potentiality for contributing to turgor, even though its specific hydrational potentiality was not as great as that of some of the other ions listed. Previously it had been noted that viewed with respect to metabolism the hydrated Na^+ ion had a modulated mobility intermediate between that of the relatively rapid K^+ ion and the sluggish movement of the extensively hydrated Li^+ ion. Previously, also it had been noted that hydrated sodium chloride possessed a retentive ability for its H_2O units which was of an order comparable with the dehydrating potentialities of the atmosphere,

Table 2. Calculated volumetric increases occasioned by the complete
hydration of the indicated element ions.

Element Ion	Wa	Va	H	Wh	Vh	$\dfrac{Vh}{Va}$	Per Cent Increase in Volume by Hydration
Ca^{++}	44	22	1	62	35	1.6	60
K^+	40	20	3	94	66	3.3	230
Cl^-	32	16	7	158	131	8.2	720
Mg^{++}	28	14	9	190	165	11.8	1080
Na^+	24	12	11	222	200	16.6	1560
Fe^{+++}	58	29	17	361	314	10.8	980
Cu^{++}	62	31	15	332	280	9.0	800
Zn^{++}	64	32	14	316	263	8.2	720

Wa = weight of anhydrous ion: Va = volume of anhydrous ion: H = number
of H_2O units potentially held in hydration: Wh = weight of hydrated ion:
Vh = volume of hydrated ion.

Collectively these considerations prompted the easy speculation that
the involved attributes might have some physiological relationship
to the rather general incidence of sodium chloride in the diet of
animals.

INDIGENOUS RADICALS

In the area of chemistry the word "radical" has denoted a group
of chemically-bound atoms which acted like a single atom with a
valence. The characterization carried no implications regarding
the stability of the involved association, but it has seemed advisable
here to point out some developments outside the area of conventional
chemical science. When researches on osmosis supplied supplementary data which documented the integrity of the description
of periodic hydrational potentiality, they also supplied a basis for
the appraisal of the behavior of solute radicals in osmosis. It then
became evident that some specific radicals involving nitrogen

linkages remained intact in osmosis and possessed potentialities for hydration precisely those prescribed by the description of periodic hydrational potentiality for element ions of the same ionic weight. It also became evident that many common radicals were completely dissociated in osmosis and that the element ions thus released became completely hydrated. The situation was complicated further by the discovery that in aqueous solutions of copper sulfate no radicals were present and that when solutions containing nitrate radicals or ammonium radicals were combined with copper sulfate and the mixtures were subjected to osmosis there took place a complete dissociation of the nitrogen-containing radicals followed by a complete hydration of the resultant element ions. Yet when some carbohydrates were subjected to osmosis there was evidenced a simple disruption of the carbon-to-carbon linkages and a subsequent complete hydration of the resultant COH radicals. Radicals remained subject to the characterization given by chemical science but it was obvious that within protoplasm their life was full of adventurous uncertainty.

Naturally it was recognized that with radicals, as with element ions, it was quite impossible to select a group whose members were common associates in the nutrition of plants and animals. Yet one could perhaps select a group whose members might be considered as representative of the indigenous radicals which might take part in nutrition. The following radicals were chosen to comprise such a group : SO_4^{--}, PO_4^{---}, HSO_4^{-}, HPO_4^{--}, CO_3^{--}, $H_2PO_4^{-}$, HCO_3^{-}. Repeatedly it had been made evident that periodic hydrational potentialities were conditioned solely by ionic weight : that an intact radical would have precisely the attraction for H_2O^{-} units which would be prescribed for an element ion of the same weight by the description of periodic hydrational potentiality. On this account it was of interest to calculate the effect of hydration on the volumes of the listed radicals considered as intact units. The results obtained have been given in Table 3.

Table 3. Calculated volumetric increases occasioned by the complete
hydration of the indicated radicals

Radical	Wa	Va	H	Wh	Vh	$\dfrac{Vh}{Va}$	Per Cent Increase in volume by Hydration
SO_4^{--}	92	46	23	506	428	9.8	880
PO_4^{---}	88	44	2	124	72	1.6	60
HSO_4^{-}	96	48	21	474	394	8.2	720
HPO_4^{--}	92	46	23	506	428	9.8	880
CO_3^{--}	56	28	18	380	331	11.8	1080
$H_2PO_4^{-}$	96	48	21	474	394	8.2	720
HCO_3^{-}	60	30	16	348	297	9.9	890

Legend as in Table 2.

A comparison of the data given in Tables 2 and 3 indicated that
the volumetric increases effected by the hydration of the involved
element ions and radicals were of the same order. For the most
part the derived values previously had been validated by the data
of specific gravity. In substantial measure the major sphere of
usefulness for such data as were given in Table 2 seemed to be
restricted to considerations of specific gravity, since in conjunction
with protoplasmic metabolism osmosis mediated the disruption of
all of the chemical bonds present in the listed radicals and effected
the complete hydration of all of the element ions thus released.
On this account it became of interest to calculate for the radicals
given in Table 3 the increases in volume effected by osmosis. The
results obtained in such a project were assembled to comprise
Table 4.

A comparison of the data of Tables 2, 3 and 4 made it quite evident
that the osmotic dissociation of solute radicals and the allied hydration of the resultant element ions brought about appreciably greater
increases in volume than those brought about by element ions as
represented in the data of Table 2 or by intact radicals as represented

Table 4. Calculated volumetric increases occasioned by complete dissociation of the radicals and complete hydration of the resultant element ions

Radical	Element Ions	Wa	Va	H	Wh	Vh	TVa	TVh	TVh/TVa	Per Cent Increase in Volume
SO₄	S⁺⁶	44	22	1	62	36	46	1260	27.4	2640
	O⁻⁻	12	6	17	318	306				
PO₄	P⁺⁵	40	20	3	94	66	44	1290	29.3	2830
	O⁻⁻	12	6	17	318	306				
HSO₄	H⁺	4	2	21	382	378	48	1538	32.0	3100
	S⁺⁶	44	22	1	62	36				
	O⁻⁻	12	6	17	318	306				
HPO₄	H⁺	4	2	21	382	378	46	1668	36.2	3520
	P⁺⁵	40	20	3	94	66				
	O⁻⁻	12	6	17	318	306				
CO₃	C⁺⁴	20	10	13	254	236	28	1154	41.2	4020
	O⁻⁻	12	6	17	318	306				
H₂PO₄	H⁺	4	2	21	382	378	48	2046	42.6	4160
	P⁺⁵	40	20	3	94	66				
	O⁻⁻	12	6	17	318	306				
HCO₃	H⁺	4	2	21	382	378	30	1532	51.1	5010
	C⁺⁴	20	10	13	254	236				
	O⁻⁻	12	6	17	318	306				

TVa = total volume of anhydrous ions; TVh = total volume of hydrated ions.   Other symbols as in Table 1.

by the data of Table 3. Moreover, for the most part the data of Table 4 had been documented as valid by previously published data on the osmotic behavior patterns of aqueous solutions of sulfates, phosphates and carbonates. There was much assurance, therefore, that indigenous radicals in conjunction with osmotic activity potentially played a substantial role in the development of hydrostatic pressure within vacuolated cells.

In the area of plant physiology and with reference to land plants directly exposed to the atmosphere it was obvious that the foregoing considerations were of primary interest in relation to the basal portions of plants : to inorganic solutes of the soil solution. Attention was next directed to the apical portions of plants : to organic solutes elaborated in the leaves.

PHOTOSYNTHATES

The results obtained in researches on the osmotic behavior of organic solutes and published in the book to which reference has been made indicated that in general carbon-to-carbon chemical linkages were subject to disruption in osmosis in a manner analogous to the description of the chemical linkages in many inorganic radicals. Linkages involving nitrogen were evidenced as less subject to disruption. When linkages were broken the resultant ionic units became completely hydrated, as was the case with disrupted inorganic radicals.

There was one seemingly very important difference between the osmotic behavior patterns of inorganic and organic solute ionic complexes. With the inorganic ions osmosis, except when nitrogen was present, effected the complete dissociation of the unit into element ions,—as was represented in the data of Table 3. With the organic ions, on the other hand, the osmotic disruption of carbon-to-carbon linkages was not followed by a complete dissociation but by the formation of subsidiary radicals containing, for most of the substances studied, one atom each of carbon, hydrogen and oxygen.

In solutes containing equal numbers of carbon atoms osmosis evidenced the formation of equal numbers of positive and negative subsidiary radicals. In all instances the pattern of osmotic behavior evidenced the disruption of carbon-to-carbon linkages and the complete hydration of the resultant ionic solute units.

The indicated developments posed some challenging questions. When it was found that the NO_3^- and NH_4^+ radicals remained intact in simple osmosis and hydrated in the manner prescribed for element ions of the same weight it was realized that in the protoplasmic metabolism of plants a breakdown of these radicals had to take place because neither radical entered intact into the composition of any known organic compound. Much later at least one mechanism of breakdown was discovered. With respect to the subsidiary COH units evidenced as released in the osmotic partial dissociation of numerous solute carbohydrates, however, the situation remained obscure. On the one hand it was possible that these units might enter directly into protoplasmic re-synthesis : the occurrence of such units in structural formulas for organic compounds was commonplace. On the other hand there were numerous structural formulas for organic compounds in which the elements, though present, were not thus associated.

With respect to turgor it was to be recognized that an osmotic disruption of the carbon-to-carbon linkages with an attendant hydration of the resultant ionic COH radicals would result in volumetric increases of an order comparable with those given in Tables 2 and 3. A complete dissociation of the COH radicals, followed by a complete hydration of the resultant element ions, would result in volumetric increases exceeding those given in Table 4.

To illustrate these observed and projected developments the volumetric increases for glucose were calculated. The results obtained have been given in Table 5.

In the data given in Table 5 the glucose molecule follows the pattern of the radiately symmetrical sucrose molecule given in

'Behavior Patterns of Hydration'. Structurally the molecule was subject to representation as cyclic with single bonds and alternate positive and negative carbon sub-ionic atoms. In aqueous solution the glucose thus represented was evidenced by specific gravity as remaining intact and a- being hydrated precisely as prescribed by the description of periodic hydrational potentiality. The characterization of this state was given in the upper portion of Table 5.

The osmotic behavior of glucose was such as to evidence a di-ruption of the carbon-to-carbon linkages and a complete hydration of the resultant COH radicals. There were at least three types of COH radicals possible as the result of the breaking of the carbon linkages, ranging from univalent to trivalent units. Insofar as osmotic behavior was concerned the average effect on volumetric

Table 5.   Calculated volumetric increases attributable to hydration for glucose in each of three projected states

Solute	Ions	Wa	Va	H	Vh	Vh	$\dfrac{Vh}{Va}$	%Increase in volume by Hydration
glucose as $C_6H_6O_6$	$C_6H_6O_6{}^{-3-3}$	180	90	2	216	112	1.24	24
glucose as $C_6H_6O_6$	COH^{+3} (3)	36	18	5	126	98	9.93	893
	COH^{-3} (3)	24	12	11	222	200		
glucose as $C_6H_6O_6$	C^{+4} (3)	20	10	13	254	236	64.93	6393
	C^{+4} (3)	4	2	21	382	378		
	O^{-} (3)	20	10	13	254	236		
	O^{-} (3)	12	6	17	318	306		
	H^{-} (3)	4	2	21	382	378		
	H^{-} (3)	0	0	23	414	414		

increa-es was the same. As represented in Table 5, central section, these COH units with a valence of three possessed hydrational potentialities of an order comparable with the hydrational potentialities exhibited by the atmosphere. At least theoretically the presence of hydrated COH units in the vacuoles of plant cells would confer upon them a measure of ability to maintain turgor.

In table 5 the data of the lower section represented a purely hypothetical extrapolation of the data of Table 4 involving a complete dissociation of the organic radicals. It was of interest that the indicated behavior patterns for all of the element ions listed had been documented previously. A recent paper on chlorophyll evidenced a molecular composition which did not include COH units and therefore constituted a basis for the viewpoint that protoplasmic synthesis began with atomic ions. Yet for the present the mechanism of dissociation for the indicated COH units, if it takes place, remains obscure.

ASPECTS OF WATER LOSS

The obvious conclusion drawn from the foregoing considerations was that the hydrostatic pressure which mediated turgor was occasioned by the increases in the volumes of solute ions—increases attributable to their hydration as brought about by osmosis. With respect to the loss of water by plants exposed to the atmosphere it became of interest to briefly review the behavior patterns associated with osmosis as given in "Behavior Patterns of Hydration." As therein described and documented, osmotically mediated increases in solution volume were effected exclusively by the hydration of solute ions : there was no increase attributable to free aqueous solvent. When osmotic activity was initiated from a situation in which all of the solute ions were on one side of a membrane, the increase in solution volume was evidenced as attributable to the hydration of precisely half the number of ions originally present, the other equal number of ions having been extruded from the other side of the membrane into the adjoining solvent. Subsequent to the completion of osmotic activity, in the event that unequal volumes of solution were present on opposite sides of the membrane, a diffusional adjustment took place which brought about a uniform concentration of solute throughout all involved solutions. This adjustment was evidenced as having involved the re-entry of solute ions into the membrane in an anhydrous state and their departure at the other interface in a hydrated state. Under laboratory conditions the magnitude of volumetric change involved in the post osmosis adjustment was negligible except under conditions involving high concentrations of solute.

In plants directly exposed to the atmosphere the introduction of the foregoing description of osmosis involved situations not present in the laboratory. Instead of one osmometer there were many cells potentially acting as osmometers. Some of these cells were internal and not directly exposed to the atmosphere. Other cells were exposed to the atmosphere and were thereby subject to water loss. All cells were adnate to a greater or lesser extent. Under these conditions a loss of water from any cell obviously would increase the concentration of solute and thereby initiate osmotic activity in relation to adjoining cells of lesser solute concentration. On the other hand in roots at any projected state of osmotic equilibrium the adjacent external liquid would contain extruded solutes. Upon the incidence of any decrease in the concentration of these external solutes, such as might be brought about by rain, osmotic activity would be initiated. Conversely, any loss of water from root cells exposed to the atmosphere would initiate an osmotically mediated withdrawal of water from adjacent internal cells.

As thus projected for plants of land habitats the creation of any imbalance with respect to turgor would be conditioned by the relative rates of water loss and osmotic activity, correlated with the ultimate accessibility of water. The removal of solutes for synthesis would be expected to involve the anhydrous state and have a minor influence on turgor, compensated by soluble products of synthesis.

The foregoing appraisal brought forward the interesting question as to whether or not the hydrational H_2O units associated with solute ions following osmosis commonly were subject to removal in conjunction with the water loss from plants attributable to the atmosphere. What seemed to be a satisfying answer to the question was supplied by the results obtained in researches by FITTING. These researches had made use of plasmolysis as an index of solute concentrations in plants and solutions of potassium nitrate were used as a reference standard. About a hundred species of plants were studied and these included plants of desert and seaside habitats. The results obtained evidenced a wide variation in the behavior of different species of plants and of groups of plants of different habitats. From the concentrations of potassium nitrate necessary to incite plasmolysis it was possible to calculate the amounts of free solvent which had been present. From these values it was calculated that plasmolysis had involved free solvent over a range extending from about 5 per cent to about 80 per cent of the free solvent present. Under these circumstances it seemed justifiable to conclude that commonly the vacuolar liquid always contained some free solvent, and that the amount of this free solvent in substantial measure conditioned susceptibility to loss of turgor. The imbalance created by loss of water and the attendant concentration of solute was subject to appraisal as a development which conferred osmotic potentiality,—a persistent potentiality whose functioning was conditioned by such factors as the availability of water, the extensibility of cell walls and temperature. The condition became subject to possible synonymy with the abstruse attribute denoted by such terms as "suction tension" and "diffusion pressure deficit."

DISCUSSION

For many years the apparent state of independence characterizing solutes in many aqueous solutions—a situation memorialized in the Kohlrausch law of the independent migration of ions—has prompted the projection for these units behavior patterns allied to the gas laws. Included among these patterns was the hypothetical concept that if these solute units suddenly were freed from their aqueous solvent they would act like gases and exert an outwardly-radiating pressure conditioned by concentration and temperature. The concept led to the use of the word "atmosphere" as a measure of solute unit concentration, but it failed to take into account the ionic nature of the solute units and the consequent existence of potentialities for hydration. With a recognition of the existence of different and characteristic forces of attraction between solutes and the H_2O- units of their aqueous solvent the word "atmosphere" became meaningless as an index of concentration, since the usage had involved the assumption that all solute units were of equal effectiveness. The early and concurrent failure to recognize that hydration involved the transfer of H_2O- units from solvent to solute led to the erroneous interpretation of electrical conductivity data which begat the incomplete dissociation theory of solutions. Unfortunately the mistakes of the past continued to persist like a fog, obscurring the true beauty of natural order.

In the present paper there has been no attempt or desire to document the principles underlying the data of the four tables : for such material the subtended references may be consulted. The development here reported is the fact that from these principles there has come evidence that the osmotically mediated hydration of solutes plays a primary and major role in the creation and maintenance of turgor. In a supplementary way the results obtained by FITTING supplied evidence that land plants under a wide range of environmental conditions maintained free aqueous solvent in their tissues and that ordinarily as "bound water" the H_2O- units in the hydration of solutes was not removed by the dehydrating forces of the atmosphere.

SUMMARY

In accordance with principles whose integrity had been thoroughly documented in previous publications the volumetric increases attendant upon the hydration of representative solutes were calculated. These increases were found to be sufficiently great to lead to the conclusion that the osmotic mediation of solute ion hydration was the primary and major process in the initiation and maintenance of turgor.

LITERATURE CITED

1. Fitting, Hans (1911). Die Wasserversorgung und die osmotischen Druckverhaltnisse der Wustenplanzen. Zeitsch. Bot. 3:209-275.
2. Flint, Lewis H. (1963). Behavior Patterns of Hydration. New Delhi, India.
3. —(1963). Concerning Chlorophyll. Advancing Frontiers of Plant Sciences 6:1-26. New Delhi, India.
4. — (1963). Concerning Color and Hydration. Advancing Frontiers of Plant Sciences 7:45-52. New Delhi, India.

CHAPTER 11. CONCERNING NITROGEN AS A METABOLITE

In the preceding volume dealing with the behavior patterns of hydration the considerations involving nitrogen for the most part were confined to nitrogen as a component of the ammonium and nitrate radicals. In each of these radicals the sub-ionic nitrogen atoms were evidenced as positively charged and pentavalent. When released from the radicals they had ionic weights of 24 and took on the 11 H_2O^- units prescribed for ions of that weight by the description of periodic hydrational potentiality. From these considerations one might venture that the N^{+5} ions had a prominent role in protoplasmic metabolism, but to remain content with so abstract a generalization would be to ignore an area of fascinating challenge.

As though in keeping with the dawn of the space age it has seemed exciting to consider the earth's atmosphere as a sea of nitrogen and earth's organisms for the most part as living at the bottom of this sea. Allied with this concept has been the thought of nitrogen as neutral and inert, of oxygen as unstable and heavier, tending to sink, and of water vapor as spirited and lighter, tending to float. Beyond the poetic charm of an imagined tranquility the characterization had romance, for out of the water below the atmospheric sea there came the organic world of baffling complexity, of surpassing beauty, of infinite potentiality, of continuous challenge. The process involved for nitrogen atoms adventurous departures from their quiet roles as components of neutral molecules in the earth's envelope. These departures were to be appraised for the present as largely mysterious and for the research of tomorrow. Nevertheless there was pleasure to be found in endeavoring to follow some of their simplest patterns of behavior.

The data pertaining to the specific gravity of aqueous solutions of nitrates, as given in Table 4 of the preceding volume, evidenced that the nitrate radical had remained intact and had taken on the 16 H_2O^- units in hydration which had been prescribed by the description of periodic hydrational potentiality for an ion of weight 60. The data pertaining to the osmotic behavior of aqueous solutions of nitrates, as given in Table 10 of the preceding volume, evidenced that the nitrate radical had remained intact and had taken on the 16 H_2O^- units which similarly had been prescribed. In sharp contrast, the data of Tables 15 and 16 of the preceding volume evidenced that in a series of aqueous solutions involving a variety a nitrogen-free radicals osmosis brought about a complete dissociation of the radicals. Thus it was made evident that the chemical bonding potentialities of nitrogen were greater than those of most other elements commonly rèpresented in solute radicals. This was not new knowledge ; chemical reactions and the more difficult digestibility of proteins as compared with carbohydrates had indicated the same thing. However, it represented confirmatory evidence from an entirely different background. The ultimate dissociation of both the nitrate and ammonium radicals was discussed in Chapter 8 of the preceding volume, and will be considered further in Chapter 16 of this volume. The important consideration at this point, however, is the evidenced superior bonding potentiality of the involved pentavalent nitrogen ions. In what might be designated as the passive state these ions have positive charges, thus being unable to supply electrons readily as energy. In contrast, in what might be designated as the potentially active state these ions hàve negative charges, thus being able to supply electrons as energy. It was fortuitous circumstances that four of these pentavalent N^{-5} ions cooperated to establish the basic strength of the chlorophyll molecule through bonding potentiality and to mediate the dynamic role of the chlorophyll molecule through electron activity in association with resonant subionic atoms of oxygen. These developments were discussed at some length in the preceding Chapter 8 of the present volume.

Attention was directed to the behavior patterns of trivalent nitrogen ions. Chapter 17 of the preceding volume had included some results obtained in studies of urea, a compound in which trivalent sub-ionic nitrogen atoms were associated with hydrogen, oxygen and carbon. The results evidenced a superior bonding potentiality for nitrogen. Although dissociative processes were indicated in water, in simple osmosis and in osmosis involving mixtures of urea and copper sulfate, the bondages between nitrogen and hydrogen remained unbroken.

The chemical formula for ammonia, NH_3, suggested an involvement of trivalent nitrogen and prompted an inquiry into aqueous solutions of ammonium hydroxide. This inquiry led into real trouble : aqueous solutions had specific gravity values less than 1.0, and this situation was not easy to correlate with the density formula $d = 1.0 + \frac{WA}{WH}$, a formula discussed and used extensively in the preceding volume. The real trouble appeared to involve the attribute of volatility. Although the behavior patterns of individual gases under specific or standard conditions formed the basis of the gas laws renowned for mathematical nicety, neither the relationships of mixed gases not the relationships of gases to an aqueous solvent had been characterized with satisfaction. A suggestive introduction to interdiffusion was given in chapter 11 of the preceding volume. It seemed important to undertake to relate the description of periodic hydrational potentiality to the solubility of gases in water, and this project will be initiated at a later point in this chapter.

With respect to aqueous solutions of ammonium hydroxide it appeared that water effected an anticipated dissociation of the material into NH_4^+ and OH^- radicals and that these radicals then hydrated to the precise extent prescribed by the description of periodic hydrational potentiality. Up to this point the indicated developments were to be considered quite commonplace, but to account for specific gravity values less than 1.0 it seemed necessary to resort to speculation. The following reaction with the solvent was projected.

$$2NH_4^+ + 2OH^- + H_2O \longrightarrow (NH_3)_2{}^{+-}.\ 3H_2O$$

The reaction with H_2O at this point was interpreted as having reduced the maximum purity to a calculated 98.2%*. The $(NH_3)_2{}^{+-}$ non-electrolytic radical had a calculated weight of 40, and the prescribed hydrational potentiality of such a radical was $3H_2O^-$ units. The hydrated radicals thus would have a weight of 94 and density of $1.0 + \frac{WA}{WH}$, $1.0 + \frac{40}{94}$, or 1.4255. On this basis a liter of these radicals would weigh 1425.5 grams, of which 607 grams would represent the weight of the $(NH_3)_2{}^{+-}$ moiety. Yet on account of the inclusion of the H_2O unit in the reaction, as indicated above, a composition projected as 100% pure would actually be 98.2% pure. In conjunction with the taking of observational data for the specific gravity of aqueous solutions of ammonium hydroxide there naturally was no recognition of the projected developments. By making allowances for the indicated circumstances it was found that calculated and observed specific gravity values could be brought into satisfactory agreement, as made evident in the data of the table at page 103.

The data of the foregoing table differed from those previously considered by virtue of the suggested volatility. Ordinarily the specific gravity of the aqueous solutions would be subject to derivation as the sums of the Column 3 and Column 6 values divided by 1000. Here there was an involvement of the sums of the Column 2 and Column 6 values divided by 1000.

The structural formula for the projected $(NH_3)_2{}^{+-}$ non-electrolytic radical was as follows.

---

*Weight of 2 NH_4^+ radicals hydrated = 444 : weight of 2OH$^-$ radicals hydrated = 572 : total = 1016 : 1016 + 18 = 1034 : 1016 ÷ 1034 = .982.

Table 6. Data affording a Comparison of Calculated and Observed Values for the Specific Gravity of Aqueous Solutions of Ammonium Hydroxide

"% NH$_3$"	Wa (NH$_3$)$_2$	Wh (NH$_3$)$_2$	Vh (NH$_3$)$_2$	Vh (NH$_3$)$_2$	Adjusted Ml Solvent	Sp. Gr. as Col. 2 + Col. 6 ÷ 1000	Observed Sp. Gr.
Col. 1	Col. 2	Col. 3	Col. 4	Col. 5	Col. 6	Col. 7	Col. 8
100	607	1425.5	1000		0	.607	
"100"=98.2	596	1400	982	982	18	.614	.618
"95"	576	1355	950	934	66	.642	.642
"90"	546	1282	900	884	116	.662	.665
"85"	516	1213	850	834	166	.682	.688
"80"	486	1142	800	786	214	.700	.711
"30"	182	428	300	294	706	.888	.892
"22"	133.5	314	220	216	784	.9175	.9164
"14"	85	200	140	137	863	.948	.943
"10"	60.7	142.7	100	98.2	901.8	.9625	.9575
"6"	36.4	85.5	60	59	941	.9774	.9730
"2"	12.13	28.5	20	19.6	980.4	.9925	.9895

Legend.   Wa=Weight of anhydrous solute in gms. per liter; Wh, hydrated solute; ml=milliliters per liter.   Sp. Gr.=Specific gravity.

It was to be recognized that in conjunction with the indicated structural formula for $(NH_3)_2{}^{+-}$ the gain or loss of two electrons potentially would result in the release of neutral NH_3 units. In a venturesome speculative way there also was the suggestion that the presence of sub-ionic H^- atoms might be a contributory factor in the volatility. By prescription these units as free atomic ions would have zero weight, a density of 1.0 referred to water, and would be buoyed up by a force equal to the weight of the water displaced. Thus in water they would be weightless. From the density formula they would be subject to appraisal as weightless in water irrespective of the satisfaction of hydrational potentiality, and such a projected relationship appears to offer a natural explanation of the unusual behavior pattern represented in the data of the preceding table. The three sub-ionic hydrogen atoms in the acetate radical bear negative charges and may well be similarly projected as contributory to volatility. Indirectly there is the suggestion that water may contain sub-ionic H^- atoms. Quite definitely water, through its evidenced ability to supply H_2O^- hydrational units normally held, has a composition more complex than that represented by the H_2O formula.

It was a distinct handicap in the development of biological science that physics and chemistry failed to supply a satisfactory description of hydrational potentiality. It is presently a distinct handicap that no satisfactory description of the behavior patterns of gases in water has been forthcoming. Lacking such a description, there is the precarious recourse to projection. The fact of solubility suggests an ionic state. The low solubility suggests a reaction with water to effect a dissociation of H_2O units and a hydration of the ions thus released. Without the description of periodic hydrational potentiality these items scarcely would supply a basis for speculation. With the description the following steps are projected as of interest.

1. Nitrogen gas as N_2 in water in an anhydrous state has a density of 2. The weight of one gram molecule would be 28 grams

and the volume would be 14 ml.

2. In water nitrogen gas becomes the duplex double-bonded ion shown in Figure 9 A.

3. Each free charge completely dissociates one H_2O unit.

4. All ions hydrate to the precise extent prescribed by the description of periodic hydrational potentiality.

5. The hydrated ions unite in chemical bondage to form the neutral solute molecule indicated in Figure 9 B.

6. The volume of the neutral solute molecule indicated in Figure 9 B was subject to calculation from the hydration and density principles.

Two procedures are available at this time. One may use the density formula, $d = 1.0 + \frac{WA}{WH}$, repeatedly validated in this book and its predecessor, to determine the total volume on a gram molecular basis. Alternatively one may add the volumes of the component sub-ionic atoms, using the reference data of Table 8. page 49 and 50, of the book "Behavior Patterns of Hydration". In each case the amount soluble in a liter of water may then be determined by ratio. The results given by the two procedures are not identical, but close. The latter procedure appears less involved,—provided the reference data are at hand.

From this volume the amount of gas soluble in a liter of water was subject to calculation.

Calculations for volume of neutral solute molecule.

Ions	Wa	H	Wh	u	Total Wa	Total Wh
H^+	4	21	382	12	48	4584
H^-	0	23	414	12	0	4968
O^{+3}	22	12	238	6	132	1428
O^{-3}	10	18	334	6	60	2004
N_2^{+6-6}	28	9	190	1	28	190
					268	13174

$$\frac{268}{13174} = .02033 \qquad \frac{13174}{1.02033} = 12900$$

Calculations for amount soluble in one liter of water.

$$14 : 12900 : : X : 1000 \qquad X = 1.083$$
$$1.083 \times 22.4 = 24.3 \text{ ml}$$

Published value : 23.82 ml.

The configuration represented in Figure 9B surely appeared to be a fantastic figment of the imagination. Nitrogen with a valence of eight has not been nurtured within the cradle of chemical science and might well expect to be disowned. Yet the seeming fantasy just might be an inorganic preview of the transcendent complexity of sub-ionic atoms assembled to initiate and improve the organic on the road to protoplasm. Be that as it may, the configuration does more than exemplify correlation with the solubility of nitrogen gas in water. It represents a pattern followed in studies of the behavior of oxygen gas in water, reported in Chapter 12, and in studies of the behavior of hydrogen gas in water, reported in Chapter 13.

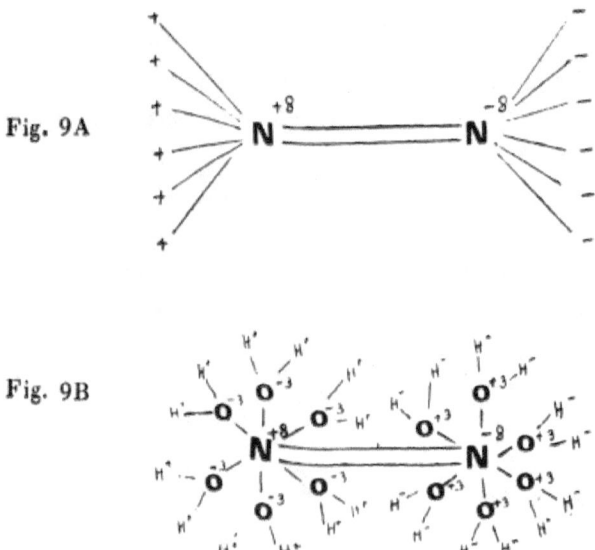

Fig. 9A

Fig. 9B

CHAPTER 12. CONCERNING OXYGEN AS A METABOLITE

With the development of the description of periodic hydrational potentiality it became the more readily understandable that water was the universal flux for metabolism. Until recently, however, the application of the description for the most part had involved what might have been designated as the more simple solutes in aqueous solutions. The monographic treatment of the description in the book "Behavior Patterns of Hydration" had included evidence that gases when ionized attained the prescribed hydrational potentialities, but in general the inter-relationships of gases and water had seemed discouragingly enigmatic.

During the course of a study of nitrogen as a metabolite the outlook changed abruptly when it was discovered that nitrogen gas appeared to react with water to form an extensively hydrated neutral solute unit. This development carried the firm implication that the solubility of gases in water, like the solubility of solids in water, involved both ionization and hydration. There was thus at hand an untried and intriguing avenue of approach to the intimately vital behavior patterns of metabolic oxygen.

IONIC OXYGEN ATOMS IN WATER

At the time when the Kohlrausch law of the independent migration of solute ions was first projected there were abundant electrical conductivity data to mediate and support it ; but gradually it came to be recognized that apparent departures from the law took place not only with increase in solute concentration but also with specific associations, even at low concentrations. At least some of these apparent departures were attributable to such factors as hydration and chemical changes which modified the nature of the solute units. Some apparent departures involving ionic oxygen atoms were selected for special study.

In an earlier publication entitled "Behavior Patterns of Hydration" the specific gravity of aqueous solutions was used rather extensively in the validation of a description of periodic hydrational potentiality. During the course of the involved research it was noted that aqueous solutions of copper sulfate were exceptional in that the specific gravity values were not successfully predictable from the assumption that the sulfate radical remained intact. Subsequently it was found that copper sulfate iu aqueous solution had unusual dissociative potentialities for the NH_4^+ and NO_4^+ radicals. Thereupon it developed that the specific gravity data for aqueous solutions of copper sulfate evidenced the complete dissociation of the assumed sulfate radical and the complete hydration of the resultant atomic ions. Since these atomic ions included oxygen it seemed appropriate to cite the specific gravity data at this point. These data were assembled to comprise Table 7.

In the data of Table 7 it was made apparent that the observed specific gravity values, taken from widely published tables, were subject to correlation with ionic attributes prescribed by the description of periodic hydrational potentiality when complete dissociation of assumed sulfate radicals was projected. The same complete dissociation was evidenced also in an entirely different approach, described in a previous chapter concerned with color and hydration. The osmotic behavior of copper sulfate in aqueous solution evidenced a complete dissociation of any assumed sulfate radicals, but this was in no sense distinctive, since osmosis was evidenced as effective in bringing about the complete dissociation of all sulfate radicals, including sulfates whose specific gravity attested the presence of intact sulfate radicals before osmosis.

From the foregoing considerations it appeared to be true that ionic solute copper as the bivalent Cu^{++} ion had the ability to bring about the complete dissociation of *any solute sulfate radical* and thus to release ionic oxygen. When released in such a manner or

Table 7. Specific gravity of CuSO₄ in aqueous solution as evidence of the presence of ionic oxygen atoms.

Ions	Wa	H	Wh	Total Wa	Total Wh	$\frac{WA}{WH}$	Total Wh ÷ 1 + $\frac{WA}{WH}$	Calculated ml solvent
Cu⁺⁺	62	15	332					
S⁺⁶	44	1	62					
O⁻⁻	12	17	318					
O⁻⁻	12	17	318					
O⁻⁻	12	17	318					
O⁻⁻	12	17	318					
				154 (1.0 molar solution)	1666	.0926	1523 (Volume Solute)	·523 Calc. Sp. gr. 1.143

% CuSO₄	Gms. CuSO₄ per liter	Calc. gms hyd. CuSO₄	Calc. ml hyd. CuSO₄	Calc. ml Solvent per liter	Calc. gms Solute + Solvent ÷ 1000	Obs. Sp. Gravity
2.557	26.22	283	259	741	1.024	1.0254
5.114	53.78	580	530	470	1.050	1.0516
7.671	82.73	894	818	182	1.076	1.0785
8.311	90.20	973	890	110	1.083	1.0854
9.589	105.4	1140	1042	-42	1.098	1.0993
11.51	129.0	1390	1280	-280	1.110	1.1208
14.06	161.7	1745	1595	-595	1.150	1.1501
17.26	205.4	2220	2030	-1030	1.190	1.1898
19.18	232.9	2510	2290	-1290	1.220	1.2146

when released through osmotically mediated dissociation the atomic oxygen uniformly was evidenced as two-valent negative, and complete hydration had always followed release. The volumetric increases which characterized osmotic behavior had been such as to evidence the acquisition of $17H_2O^-$ units in hydration, commensurate with the weight 12 prescribed by the description of periodic hydrational potentiality for O^{--} ions. As completely hydrated O^{--} ions the calculated and evidenced weight was 318, -a weight which quite obviously imposed restraint upon mobility. The calculated and evidenced density of the hydrated ions was 1.0377, -a value which yielded 306 as a correlated measure of volume. Repeatedly it had been evidenced that completely hydrated ions were capable of entering into chemical combinations without a loss of their associated H_2O^- units. There was evidence that associated units or sub-ionic components of H_2O^- units catalyzed chemical reactions. In a previous Chapter dealing with chlorophyll it was pointed out that the modification of chlorophyll a to form b and the modification of carotin to form xanthophyll involved the addition of sub-ionic components of H_2O and in consequence carried the suggestion of an adaptation accompanying and helping to mediate the transition from an aqueous to an aerial environment. Finally, in the tabulated osmotic data included within the book entitled "Behavior Patterns of Hydration", it had been made very clear that osmosis brought about a complete dissociation of such solute radicals as sulfates, phosphates and carbonates, and that the constituent atomic ions thus released became completely hydrated. Collectively the foregoing considerations were appraised as having prescribed for hydrated O^{--} ions a prominent role in metabolism. Inasmuch as hydrated ions repeatedly had been evidenced as entering absorptive membranes in an anhydrous state, it followed that within protoplasm O^{--} ions could be present in either the anhydrous or the hydrated state, the particular state having been conditioned by such factors as the availability of H_2O^- units and the nature of the associated ions and colloids.

Inasmuch as a common incidence of O^{--} ions in water had beeu noted it was natural to assume that these ions were far more stable in water than their opposites, the O^{++} ions. Yet there was abundant evidence that O^{++} ions took part in metabolic processes. For example, there was the evidenced presence of sub-ionic O^{++} atoms in chlorophylls a and b. The resonant O^{++} sub-ionic atoms in chlorophyll a appeared to have important and dramatic roles in photosynthesis.

That six-valent oxygen took part in metabolism was evidenced by the results obtained in some experiments which involved oxygen and its chemical analogs sulfur, selenium and tellurium. In these experiments numerous short single-bud sections of sugarcane were immersed in liquids. Sections were removed at intervals and placed in moist chambers. Bud germination was used as an index of survival and recovery. The results obtained have been assembled to comprise the following Table 8.

The data given in Table 8 were interpreted as follows. The nature of the gases released constituted evidence that each of the chemical analogs had been involved in respiration. The survival and recovery data were considered to be in substantiation of this evidence, since the mobility of the listed analogs decreased from left to right in the Table. The original status of the S, Se and Te elements as sub-ionic atoms was 6-valent positive, a situation which carried the implication that 6-valent positive sub-ionic atoms of oxygen had been present in water. The presence of 2-valent negative sub-ionic analog atoms in the released gases was appraised as evidence of their antecedence as free solute ions in the respective liquids. In conjunction with osmotic dissociation the evidence implied a potential metabolic role for the oxygen in the sulfates, phosphates and carbonates. The precipitation of elemental selenium and tellurium was considered attributable to the activity of contaminant microorganisms.

Table 8.   Results obtained in experiments involving submerged buds of sugarcane stalks.

Chemical Analogs	O	S	Se	Te
Liquid in which plant Structure were Immersed	Water	$Na_2 SO_4$ in aqueous solution .05 molar	$Na_2 Se O_4$ in aqueous solution .05 molar	$Na_2 Te O_4$ in aqueous solution Saturation
Average survival in Days	8	13	21	30
Average recovery in Days	7	11	18	26
Gas Evolved	H_2O	H_2S	H_2Se	H_2Te
Deposition at end of experiment	None	None	Salmon	Black

ATMOSPHERIC OXYGEN IN WATER

The essentiality of oxygen to the respiration of most plants and animals and the importance of water to the maintenance of life appeared to be so correlated as to give emphasis to a study of the oxygen—water relationships. The investigation naturally involved the description of periodic hydrational potentiality. The investigation had to be empirical, but the results which had been obtained in a study of the behavior patterns of nitrogen gas in water were most suggestive and helpful. Ultimately a satisfying appraisal of the involved developments was obtained. The appraisal became subject to representation in items as follows.

1. O_2 gas in water becomes ionized as in the diagram.

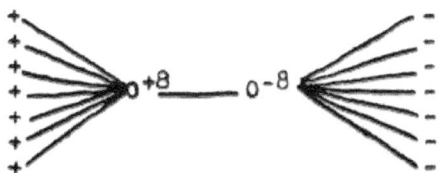

This ion was very similar to the one projected for nitrogen gas in water.

2. A complete dissociation of H_2O units by the O^{+8} moiety to yield two types of ions, H^+ and O^{-3}.

There was evidence that the dissociation potentialities of the oxygen ion were not as great as those of the nitrogen ion, since the latter had evidenced a complete dissociation of H_2O units to yield four types of ions : H , H^+, O^{-3}, and O^{+3}.

3. A complete hydration of the ions indicated in item 2.

This development followed the usual behavior pattern. The extent of the hydration was precisely as prescribed by the description of periodic hydrational potentiality.

4. A synthesis to yield a duplex ion, one part of which was extensively hydrated and one part of which was anhydrous. The anhydrous moiety as a free ion had a calculated weight of zero, a mobility calculable as infinity under the Graham

Law of Diffusion, and in situ its single bond to the hydrated moiety
was of a type which had been evidenced as subject to casy disruption.
The duplex ion became subject to representation as in the following
diagram, figure 10.

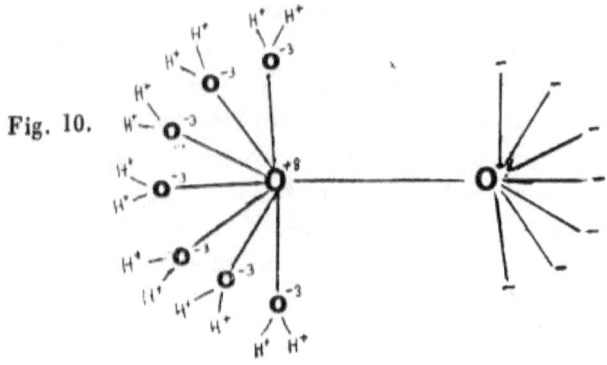

Fig. 10.

5. The volume of a gram molecule of the item illustrated in
Figure 10 was subject to calculation in the following manner.

The calculations :

Ions	Wa	H	Wh	Vh	u	Total Vh
H^+	4	21	382	365	14	5110
O^{--}	10	18	334	325	7	2275
O_2	32					16
					Total	7401

For a liter volume :

x : 1000 : : 16 : 7401

x = 2.16

For volume of gas in air :

2.16 times 22.4 = 48.4 ml.

A commonly reported value for the solubility of oxygen gas in a
liter of water = 48.89 ml.

DISCUSSION

The present chapter is a report of the results obtained in studies
of the behavior patterns of oxygen appraised from a background of

familiaity with a description of periodic hydrational potentiality. As the first such report and appraisal it was recognised that the procedures involved were in part necessarily empirical. Since the contemporary biophysical and biochemical interpretations were formulated in the absence of any satisfactorily definitive knowledge of hydration it was natural that the introduction of unfamiliar items should tend to deprecate some traditional and perhaps treasured viewpoints. With respect to oxygen a major deprecation involved the fundamental behavior pattern of a prescribed change in weight with ionization. The data involving an ionic weight of 12 for the anhydrous O^{--} ion were somewhat exceptional as representing non-osmotized aqueous solutions, although a few other solutions demonstrably attested the presence of free O^{-} ions. However, a preceding publication had included a substantial accumulation of various types of data which repeatedly had confirmed the integrity of the prescribed change in weight with ionization.

The suggestion of evidence of the presence of 8-valent sub-ionic atoms had been a surprise in the case of biatomic nitrogen gas in water, but was a reassurance in the case of biatomic oxygen gas in water. Since dissociative reactions and extensive hydration were evidenced as involved in both cases it was understandable that the solubility relations had remained enigmatic for a long period. With valences of 2 and 8 for ionic oxygen the incidence of intermediate valence values occasioned no surprise. The well-established validity of H_2O^- as an hydrational unit normally held in water rather presupposed a potential valence of 3 for oxygen.

With respect to oxygen as a mediator of respiration there was the suggestion from the reported reactions between gaseous oxygen and water that the O^{-8} moiety was the involved source of electrons as energy. The single bond of the moiety attachment seemed amenable to disruption rather easily, and as a free ion its initial weightlessness gave it a potential mobility which from the Graham Law of Diffusion was calculable as infinity in a medium offering no resistance. With all life appraised phylogentically as having arisen within an aqueous matrix it was natural to interpret adaptations to aerial or land environments on the part of plants and animals as having included provision for the maintenance of the oxygen-water reactions in the alveolar tissues of leaf mesophylls and lungs respectively. In this connection it was of interest to note that from the description of periodic hydrational potentiality and Graham's Law of Diffusion both the H- ions and O-8 ions not only were prescribed as weightless and with infinite mobility, but also shared the maximum hydrational potentiality of 23 H_2O units. The prescribod and appreciably-evidenced attributes appeared to nicely orient biology for integration with the space age.

CHAPTER 13.   CONCERNING HYDROGEN AS A METABOLITE

About a century and a half ago Prout advanced the hypothesis that all of the chemical e! ments had been constructed with units of a common substance,—Ł. drogen.   The behavior patterns of gases had impressed him,—as they might well impress anyone.   From these patterns came whole numbers as atomic weights :   C$=$12, N$=$14, O$=$16.   Why not whole numbers for the atomic weights of all the elements ? It was an intriguing question. Repeatedly chemists have stated that they would like to believe in the Prout hypothesis and, as often, various reasons have been brought forward to discredit it.   Important among these reasons was the contemporary atomic weight assigned to the basic element hydrogen.

When studies of the X-ray diffraction patterns of the chemical elements led to their characterization by specific whole numbers— the atomic numbers—a step had been taken in the direction of an acceptability for the Prout hypothesis.   For the mentioned elements the values became related as follows : carbon ; atomic number 6, atomic weight 12 ; nitrogen : atomic number 7, atomic weight 14 ; oxygen : àtomic number 8, atomic weight 16.   As thus represented the atomic weight was twice the atomic number.   Apparently in chemical science it was of no concern that the atomic weight values had been derived from the behavior patterns of gases,—a state characterized by a substantiàl measure of freedom for the individual constituent units, in sharp contrast to the solid state.   Apparently in chemical science it was of no concern that the atomic weight values had been derived by inference from the attributes of such gases as N_2, O_2, and CO_2, at a time in which atoms were appraised as solid balls of the utmost minuteness, and that actually no measurements of individual atoms had been involved.   In any event it was obvious that for consistency the atomic weight of hydrogen as element number 1 had to be 2, and this value 2 was not in keeping with the viewpoints of chemical science. The hypothesis of Prout continued to be discredited.

In 1866 Graham described a behavior pattern for gases which came to be known as the law of diffusion. The law seemed to have a kinship with Newtonian gravity and was perhaps not strictly within the province of chemistry, but Graham had included it in his popular textbook of chemistry and chemical science appeared to consider it as its own. The law characterized diffusional mobility as an inverse function of the square root of density. Avogadro had indicated that under prescribed conditions density and the weight of the individual components of gases were directly related. In a combination of the involved characterizations it followed that for the individual components mobility was inversely proporional to the square root of the weight. To an appreciable extent this combination or revised version came to be known as the law of diffusion.

In the early days of chemical science the major goal was to identify, characterize and differentiate all of the chemical elements of the earth, and one of the most critical and prized attributes of any element was its atomic weight. Innocently but unfortunately the science became obsessed with gravimetric rituals and statistics became intimate with irregular fractions. Before the arrival of devices called mass spectrographs irregular fractional atomic weight values had been installed for most of the chemical elements. After the arrival the nature of the weight values was attributed to interestingly appropriate jumbles of differing atoms having integral weights. If it ever occurred to any chemist that weight might change with ionization and that the unusual stresses imposed in mass spectrographs induced ionic diversity the light was hidden under a bushel as sheer heresy.

It was natural that in a chemical world surfeited with irregular fractional atomic weight values some apparently simple behavior patterns of gaseous hydrogen and oxygen received special attention.

It was found that when these gases were allowed to pass through tubes the hydrogen moved at a rate which was four times the rate at which the oxygen moved. In these behavior patterns there was apparent confirmation of the contemporary weight of 2 for the biatomic hydrogen molecule. The reciprocal of the square root of 2, representing the hydrogen molecule, was four times as great as the reciprocal of the square root of 32, representing the biatomic oxygen molecule. Since chemical science had derived a weight of 1.0, or thereabouts, for hydrogen atoms and a weight of 16 for oxygen atoms the observed behavior patterns of the respective gases in tubes comprised an impressive showcase for the authenticity of the gravimetric procedures. For many years the patterns were accorded prominence in text-books of chemistry.

In a report on the behavior patterns of hydration a consideration of interdiffusion, as contrasted with diffusion, clearly emphasized the fictitious nature of the conventional interpretation of the observed behavior of hydrogen and oxygen in tubes. Under the prescribed conditions with air in the tubes at the outset, the rate of interdiffusion for hydrogen was four times the rate of interdiffusion for oxygen only in the event that the molecular weight of the oxygen was 32 and the molecular weight of the hydrogen was 4. This development removed an interpretation which for a long time had served as a seemingly valid basis for discounting the theory of Prout. The development also was precisely in keeping with the corollaries of the description of periodic hydrational potentiality, —corollaries which had characterized and differentiated categories of weight, and corollaries whose validity repeatedly had been attested by the behavior patterns of hydration.

In keeping with the foregoing considerations the present survey of hydrogen as a metabolite was projected from a background of complete confidence in the integrity of the value 2 as the atomic weight of neutral hydrogen. The survey was an outgrowth of the results of previously reported surveys involving the behavior patterns of nitrogen and oxygen as metabolites.

HYDROGEN ATOMS AS SOLUTE IONS

Commonly solute H^+ ions have been most prominent in relation to considerations of acidity. In the early years of the present century solute H^+ and OH^- ions were projected as constituting an exclusive reciprocating fellowship which mediated all acid-alkaline reactivity. There were objections to the interpretation. The respective specific electrical conductivities of these ions were constants derived from studies of such solutes as HCl and $NaOH$, correlated with similar studies of solute NaCl. The electrical conductivity data for solute HCl therefore testified to the absence of OH^- ions and data for solute NaOH testified to the absence of H^+ ions. The relatively great conductivity value for the solute H^+ ion gave it prominence in relation to electrical conductivity and the wheatstone bridge endowed electrical conductivity measurement techniques with superlative sensitivity. Unfortunately again there was no essential correlation between accuracy in measurement and accuracy in interpretation. The interpretation was quite inaccurate, and at the present time generations of students have been indoctrinated with reciprocals of logarithms as indices of acid-alkaline reactivity. For an improved understanding of metabolic hydrogen a reappraisal seems quite essential.

A basic source of confusion in regard to the solute H^+ ion was the absence of a knowledge of the involved hydrational potentiality and of the relation of this potentiality to the environment. The description of periodic hydrational potentiality supplied the necessary mathematical background knowledge and prescribed for H^+ ions a weight of 4 and an ability to take on 21 H_2O units in hydration. That such an ability was a characteristic of solute H^+ ions had been evidenced in several ways : by the indicated solubility of gaseous hydrogen, by observed osmotic increases in solution volumes and by observed specific gravity values for aqueous solutions. For convenience in illustration, data on osmotic increases and specific gravity have been assembled to comprise Tables 9 and 10.

It was to be noted that in the data of Tables 9 and 10, involving aspects of volumetric increase attendant upon hydration, the assumed and evidenced weight of the anhydrous H^+ ions was 4, and the assumed and evidenced number of H_2O units, the H value, taken on in hydration was 21, as prescribed by the description of periodic hydrational potentiality. In the data of Table 9 the incidence of element ions was attributed to the readily demonstrable osmotic potentialities for dissociation. In the data of Table 10 the involved solute was one of several whose attributes in aqueous solution, independent of osmotic activity, evidenced a complete dissociation of any projected radicals.

As evidenced by the foregoing considerations it became quite clear that since all metabolic activity was subject to direct or indirect correlation with the presence and participation of vacuolar tissue having osmotic potentialities hydrated H^+ ions exercised potential roles as metabolites. Moreover, with an extensive hydration amounting to 21 H_2O units it was clear also that in the roles the hydrated H^+ ions were not entities with the dynamic attributes which commonly characterized their behavior in aqueous solutions of acids. Fortunately the situation had a simple explanation which became revealed when the pattern of behavior under electrical stress was correlated with the data of specific gravity for acids in aqueous solution.

In conjunction with the Graham law of diffusion the description of periodic hydrational potentiality permitted the calculation and comparison of relative mobility values for ions in anhydrous and in hydrated states. For convenience in the writer's studies these mobility values were calculated as the reciprocals of the square roots of the ionic weights times 10^4. For univalent ions the relative electrical conductivity values were assumed to be an order identical

with the relative mobility values. Using the observed electrical
data for the hydrated solute potassium ion K^+ as a base it became of
interest to calculate the corresponding value for the solute H^+ ion in
solutions of acids and to compare the derived value with observed
measurements. The involved considerations have been indicated in
Table 11.

Table 9. Date evidencing the osmotic behavior of the solute H^+ ion
in the indicated aqueous solutions. Condensed from Table 15 in
"Behavior Patterns of Hydration."

Solute	Assumed Ions	Wa	H	Wh	Calculated Specific Osmotic Increase in volume	Observed Specific Osmotic Increase in volume
Na_2HPO_4	Na^+	24	11	222		
	Na^+	24	11	222		
	H^+	4	21	382		
	P^{+5}	40	3	94	957.5	957
	O^{--}	12	17	318		
	O^{--}	12	17	318		
	O^{--}	12	17	318		
	O^{--}	12	17	318		
K_2HPO_4	K^+	40	3	94		
	K^+	40	3	94		
	H^+	4	21	382		
	P^{+5}	40	3	94	807.5	807
	O^{--}	12	17	318		
	O^{--}	12	17	318		
	O^{--}	12	17	318		
	O^{--}	12	17	318		

Legend : Wa = weight in the anhydrous state ; H = number of H_2O^- units
potentially held in hydration ; Wh = weight in the hydrated state.

From the data given in Table 11 it was quite obvious that within
the applied electrical field the H^+ ion had moved precisely in the
manner prescribed for an ion of weight 4. This development constituted further confirmation of the description of periodic hydrational potentiality, if such was considered desirable, but the impor-

tant thing at this point seemed to be the fact that except for the previously considered evidence for the existence of solute H^+ ions in the hydrated state there was no apparent reason for implicating the electrical stress as a dehydrating agent. With such evidence, however, it seemed important to survey the observational data on the specific gravity of acids in aqueous solution. Such a survey had been made as a portion of a previously published report on the behavior patterns of hydration. For convenience, excerpts were assembled to comprise the data of Table 12,

Table 10. Data on the specific gravity of aqueous solutions of acid sodium sulfite, $NaHSO_3$. Gms. per liter and observed specific gravity values taken from widely published tables. Symbols as in Table 9. Observed values suggest incidence of oxygen from the air.

Assumed Ions	Wa	H	Wh	$\frac{Wa}{Wh}$	Density
Na^+	24	11	222		
H^+	4	21	382		
S^{+6}	44	1	62		
O^{--}	12	17	318	.0666	1.0666
O^{--}	12	17	318		
O^{--}	12	17	318		

Gms. per liter	Gms.÷ .0666	Gms. Hydrated Solute÷ 1.0666	Calculated Ml Solvent	Calculated Specific Gravity	Observed Specific Gravity
21.46	322	302	698	1.020	1.022
37.37	560	526	474	1.034	1.038
53.65	806	757	243	1.049	1.052
69.42	1040	976	24	1.064	1.068
86.72	1300	1220	-220	1.080	1.084
104.50	1568	1470	-470	1.098	1.100

Such data as those given in Table 12 made it quite evident that aqueous solutions of many acids in the absence of electrical stress contained hydrated H^+ ions. Yet the data made it equally evident that anhydrous ions were present in such solutions, and this was in contrast to the indicated absence of anhydrous ions in aqueous solutions of many salts and sugars and in all solutions following osmosis.

Table 11.   Data evidencing the anhydrous state of the solute H^+ ion under electrical stress.

Solute Ions	Assumed State	Calculated Weight	Calculated Mobility Value	Observed Specific Ion Mobility Under Electrical Stress
K^+	hydrated	94	1031	64.5*
K^+	anhydrous	40	1581	
H^+	hydrated	832	512	
H^+	anhydrous	4	5000	

Calculations : 1031 : 5000 : : 64.5 : X   X = 313

Observations : Relative specific H^+ ion mobility under electrical stress : 313

*Reference—Creighton, 1924.   Principles and applications of electrochemistry, Vol. I.   John Wiley and Sons, New York.

Such solutions as those listed in Table 12 reacted with absorptive membranes and did not participate in osmosis.   Previous studies had yielded results which indicated that when adjustments to concentration differences took place following osmosis re-entries to the membrane were always in the anhydrous state, as were the initial entries.   The obvious dissociation potentialities of absorptive membranes evidenced the involvement of stresses comparable with those imposed by an electrical field.

Out of the foregoing considerations there came the suggestion that under such conditions of stress as those imposed by electrical fields, by the presence of anhydrous ions and by absorptive membranes solute H^+ ions became anhydrous.   Speculatively such a development precipitately would impart a dynamic potency for taking part in metabolic processes.   It would also render their patterns of behavior exceedingly rapid, transitory and elusive. As thus appraised the relatively slow and stable hydrated H^+ ions were cast in the role of reserves, the anhydrous H^+ ions serving as active participants in metabolic processes.

HYDROGEN ATOMS AS SOLUTE H^- IONS

In the foregoing appraisal of acids in aqueous solution hydrogen atoms were represented as present in the H^+ or positive ionic state. To an appreciable extent the dynamic activity of acids has appeared to overshadow alternative roles for hydrogen. In aqueous solutions atomic hydrogen often has been appraised and represented as present in the positive ionic state exclusively.

As prescribed by the description of periodic hydrational potentiality the anhydrous negative ion of hydrogen, H^-, was by far the most dynamic element on earth. Under the Graham law of diffusion its relative mobility in the absence of inter-diffusional substance was calculable as infinite,—a development which conferred a characteristic of radiation. Under the circumstances one might venture that the activity of anhydrous H^- ions had been so dynamic as to have been elusive and to have escaped appropriate recognition. Since the H-ion had a prescribed combining weight of 1.000 one might venture further that historically the ion had been involved in the derivation of the contemporary conventional atomic weight value assigned to hydrogen.

Up to the present time and in general chemical science has failed to indicate the positive or negative nature of the atomic ionic components of a great many molecules. On this account the incidence of H^- ions as sub-ionic atoms has been largely a matter of surmise. In the NH^+_4 radical, for example, the sub-ionic atom of nitrogen has been subject to projection as five-valent positive, in which case it followed that each of the four associated hydrogen atoms bore a single negative charge. Or again, in the acetate radical the three sub-ionic atoms of hydrogen were recognized as subject to

Table 12. Data evidencing the existence of hydrated H^+ ions with accompanying anhydrous ions in aqueous solutions of the indicated acids. Condensed from Table 34, "Behavior Patterns of Hydration."

Solute	Assumed Ions and Wa Values	Assumed State	Ionic Wts.	Conc. in Gms. per liter	Calculated Specific Gravity	Observed Specific Gravity	Indicated Original State of Solute
HF	H^+ = 4 F^- = 16	H A	382 16	157.95	1.055	1.053	1 H_2O
HCl	H^+ = 4 Cl^- = 32	H A	382 32	172.4	1.0765	1.0776	1 H_2O
HNO_3	H^+ = 4 NO_3^- = 60	H A	382 60	260.8	1.135	1.134	1 H_2O
HBr	H = 4 Br^- = 68	H A	382 68	1148.875	1.762	1.7675	A
H_3PO_4	H^+ = 4 H_3PO_4 = 96	H A	382 96	109.2	1.0725	1.0764	A
HI	H = 4 I^- = 104	H A	382 104	155.27	1.103	1.1091	A
H_2SeO_4	H^+ = 4 H^+ = 4 SeO_4^- = 130	H H A	382 382 130	1697.6	2.122	2.120	A
H_2SO_4	H^+ = 4 H^+ = 4 SO_4^{--} = 92	H A A	382 4 92	240.9	1.144	1.147	A

Observed values from widely published tables.

replacement by such ions as F-, Cl-, Br-, and I-, on which account there was the implication that each of the three sub-ionic atoms of hydrogen bore a single negative charge.

For the present consideration the most favourable approach to a study of the behavior of solute H- ions appeared to be that involving the complete dissociation of solute NH_4^+ radicals. In a previous report on the behavior patterns of hydration Chapter 8 dealt with the release of atomic nitrogen ions and the data of Table 19 evidenced the complete dissociation of the ammonium NH_4^+ radical. In the context such a dissociation placed emphasis on the development with respect to nitrogen in nutrition, but the data were equally significant with respect to hydrogen. The involved data have been abbreviated and reassembled to comprise Table 13.

The data given in Table 13 were interpreted as evidence that the four sub-ionic H- atoms in the NH_4^+ radical upon release possessed anhydrous weights of zero and took on 23 H_2O units in hydration, precisely as prescribed by the description of periodic hydrational potentiality. By virtue of these data the involved H- ions rather clearly attained a status as potential metabolites. These data, however, were obtained during a period directly following the discovery of osmotic potentialities for dissociation. Subsequently it was found that the specific gravity of aqueous solutions of copper sulfate evidenced a complete dissociation of the projected sulfate radical. It became of interest, therefore, to determine the specific gravity of mixed ammonium chloride and copper sulfate salts in aqueous solution. There were no available published data for such mixtures but the procedures for the determinations were simple. The results obtained have been assembled to comprise Table 14.

The data of Table 14 were noted as in agreement with those of Table 13, both sets of data being in conformity with behavior patterns prescribed by the description of periodic hydrational potentiality. It was considered of interest that the stability of the indicated hydrated H- ions appeared to be conditioned by the presence of specific associates, since commonly such ions did not appear to be present and under many conditions the ammonium radical had been evidenced as remaining intact in aqueous solution. The indicated specific associates, however, were by no means inaccessible to osmotically active tissues, on which account hydrated H- ions clearly were potentially functional as metabolites.

Table 13. Osmotic evidence of the presence of H⁻ ions. Adapted from Table 19, "Behavior Patterns of Hydration."

Solute	Assumed Ions	Wa	H	Wh	Calculated Specific Osmotic Increase in Volume	Observed Specific Osmotic Increase in Volume
	NH_4^+	24	11	222		
	Cl^-	32	7	158		
NH_4Cl	Cu^{++}	62	15	332		
	S^{+6}	44	1	62	834.5	
+	O^{--}	12	17	318		
	O^{--}	12	17	318		
$CuSO_4$	O^{--}	12	17	318		
	O^{--}	12	17	318		
	N^{+5}	24	11	222		
	H^-	0	23	414		
NH_4Cl	H^-	0	23	414		
	H^-	0	23	414		
+	H^-	0	23	414	1662.5	1662
	Cl^-	32	7	158		
$CuSO_4$	Cu^{++}	62	15	332		
	S^{+6}	44	1	62		
	O^{--}	12	17	318		
	O^{--}	12	17	318		
	O^{--}	12	17	318		
	O^{--}	12	17	318		

With respect to their function it was to be recognized that the extensively hydrated H- ions, like their counterparts the extensively hydrated H+ ions, were subject to appraisal as metabolic reserves. Like their counterpart H+ ions also, under a variety of stress condi-

Table 14. Data on specific gravity evidencing the incidence and hydrational behavior pattern of negatively-charged solute hydrogen ions.

Solutes	Assumed Ions	Wa	H	Wh	Molar Concentration	Ml Solvent per liter	Calculated Specific Gravity	Observed Specific Gravity
NH_4Cl	N^{+5}	24	11	222				
	H^-	0	23	414				
$+$	H^-	0	23	414	$\frac{M}{6}$	416	1.033	1.033
	H^-	0	23	414				
$CuSO_4$	Cl^-	32	7	158	$\frac{M}{4.5}$	124	1.050	1.050
	Cu^{++}	62	15	332				
	S^{+6}	44	1	62				
	O^{--}	12	17	318	$\frac{M}{3}$	-168	1.066	1.066
	O^{--}	12	17	318				
	O^{--}	12	17	318				
	O^{--}	12	17	318				

tions they would become anhydrous,   and be subject to projection as superlatively dynamic in relation to metabolic processes.

INTER-RELATIONSHIPS OF HYDROGEN IONS

Except under conditions of electrical stress the data relating to solute hydrated hydrogen ions in the foregoing Tàbles evidenced for such ions a substantial degree of stability in aqueous solution. On this account there was no basis for anticipating any interaction of hydrated H^+ ions and hydrated H^- ions. Yet for added assurance it became of interest to prepare an aqueous solution designed to contain both types of hydrogen ions. It was obvious that the specific gravity of such a solution readily could be calculated and that the measurement of the specific gravity of the prepared solution was a simple matter. Following a pattern indicated in preceding Tables a solution assumed to include hydrated H^+ ions and hydrated H^- ions was prepared. The specific gravity of the solution was calculated and measured. The results obtained were assembled to comprise Table 15.

The data of Table 15 were interpreted as evidence that hydrated H^+ ions and hydrated H^- ions exhibited no interaction in aqueous solution and that they therefore could exercisé contemporary roles as metabolites.

HYDROGEN GAS IN WATER

The studies of the behavior patterns of nitrogen gas in water, as reported in Chapter 11, and of oxygen gas in water, as reported in Chapter 12, yielded results which were most helpful in the interpretation of the behavior patterns of hydrogen gas in water. Both nitrogen and oxygen evidenced the presence of 8-valent element ions, 8^+ and 8^-, linked in pairs. The value 8 was subject to correlation with the outer orbital pattern of the involved elements. In the case of hydrogen the outer orbital pattern was subject to representation as 2 instead of 8, and by analogy it was projected that the

Table 15. Data evidencing the presence and stability of hydrated H^+ ions and hydrated H ions in the same aqueous solution.

Solutes	Assumed Ions	Wa	H	Wh	Molar Concentration	Ml Solvent per liter	Calculated Specific Gravity	Observed Specific Gravity
NaHSO₄	Na⁺	24	11	222				
	H⁺	4	21	382				
	S⁺⁶	44	1	62				
	O⁻⁻	12	17	318				
NH₄Cl	O⁻⁻	12	17	318	$\dfrac{M}{12}$	556	1.026	1.026
CuSO₄	O⁻⁻	12	17	318				
	O⁻⁻	12	17	318				
	N⁺⁵	24	11	222				
	H⁻	0	23	414				
	H⁻	0	23	414				
	H⁻	0	23	414	$\dfrac{M}{6}$	112	1.052	1.052
	H⁻	0	23	414				
	Cl⁻	32	7	158				
	Cu⁺⁺	62	15	332				
	S⁺³	44	1	62				
	O⁻⁻	12	17	318				
	O⁻⁻	12	17	318				
	O⁻⁻	12	17	318				
	O⁻⁻	12	17	318				

atoms in biatomic hydrogen gas would exhibit valences of 2 in aqueous solution. Quite naturally this would have appeared to be a preposterous or utterly too daring a projection had it not been for the studies of nitrogen and oxygen. The latter studies also had evidenced for every unsatisfied charge a dissociation of one solvent unit, a complete hydration of the ions thus released, and a synthesis to form an aggregate extensively hydrated non-ionic molecule. Actually these preceding patterns of behavior made it a simple matter to calculate the volume of hydrogen gas soluble in a liter of water. The steps were as follows.

1. H_2 in water $+$ $-$ H^{+2}—H^{-2}— —
2. Dissociation of $2H_2O$ units to yield the solute ions O^{-3}, H^+, H^+ and O^{+3}, H^-, H^-.
3. Complete hydration of the listed ions.
4. Aggregation of the hydrated ions to form the molecule given in figure 11.

Fig. 11

From the foregoing in conjunction with the description of periodic hydrational potentiality and the allied density relationship it was a simple matter to calculate the volume in milliliters of one gram molecule of the item 4 aggregate. It was especially simple if one used the reference table given on pages 49 and 50 of the preceding volume.

The calculations :

H^+	$Wa=4$	$Vh=365$
H^+	$Wa=4$	$Vh=365$
O^{-3}	$Wa=10$	$Vh=325$
H^-	$Wa=0$	$Vh=414$
H^-	$Wa=0$	$Vh=414$

O^{+3} $Wa = 22$ $Vh = 218$

H_2 $Wa = 4$ 2

Total Volume   2103 ml.

$X : 1000 : : 2 : 2103$ $X = .951$

For volume of gas in air :

$.951 \times 22.4 = 21.30$

Calculated ml. gas soluble in one liter
of water

A commonly reported observed value $= 21.48$ ml.

From the foregoing correlation it became evident that the behavior of the three biatomic gases N_2, O_2 and H_2 with respect to solubility in water followed a common pattern. It also was obvious that the description of periodic hydrational potentiality was an indispensable adjunct to the detection of the indicated common pattern of behavior. Collectively the results obtained in the studies of the solubility of the three biatomic gases in water seemed subject to appraisal as having documented an advance in a salient on the frontiers of knowledge.

DISCUSSION

The foregoing considerations represented solute hydrogen atoms as potential reserve metabolites which might occur in either of two ionic forms, H^+ and H^-. In the presence of adequate solvent the patterns of hydrational behavior for each of these forms were precisely those prescribed by the description of periodic hydrational potentiality. The two ionic forms differed in their weights in the anhydrous state, in the numbers of H_2O units potentially held in hydration and in their respective weights in completely hydrated states. The common incidence of H_2O units in chemical compounds did not prevent chemical activity, but with respect to hydrogen under conditions of stress sufficient to effect dehydration both types of ions were indicated as particularly dynamic potential participants in metabolic processes. Included as types of stress were absorptive membranes, extraneously applied electrical fields and internal electrical fields accruing through ion proximity, especially proximity involving anhydrous ions.

Beyond the indicated roles for hydrogen as reserves and reactive metabolites the solubility of biatomic hydrogen gas in water, taken in conjunction with previously reported studies of gaseous nitrogen and oxygen in water, suggested that the involved elements, in contrast to the solubility of salts, did not dissociate but became ionic and dissociated units of the aqueous solvent. The ions thus released, following hydration, united with the original paired sub-ionic atoms to form neutral molecular aggregates.

The indicated hydrational behavior pattern for solute H^- ions, in conforming to the pattern prescribed by the description of periodic hydrational potentiality became evidence for the prescribed zero weight of the ions in the anhydrous state. The concept of a substance of zero weight certainly was not a new one : Mendeleev in Russia and Harkins in the United States had written concerning it. Nevertheless it seemed of interest that whereas previously zero weight speculatively had been projected as an attribute of an undiscovered element, in the present instance zero weight had been both projected and evidenced as an attribute of anhydrous negatively charged atomic hydrogen. Such hydrogen, projected as a component of solar radiation, seemed to merit consideration as supplying a chemical basis for the accretionary theory of the evolution of the sun's planetary system. Such hydrogen either on contact with moisture to effect hydration or upon impact to effect the loss of an electron would acquire weight and thereupon become subject to gravitational forces. Either or both of these speculative eventualities also might be involved in the mediation of metabolic processes.

SUMMARY

A survey of the behavior patterns of solute hydrogen evidenced existence of two types of ions ; H⁺ ions and H⁻ ions. When in the completely hydrated state and accompanied by other hydrated ions both types of hydrogen ions were indicated as stable and potentially serviceable as metabolites. The evidenced characteristic degree of hydration for each type of ion was precisely as prescribed by the description of periodic hydrational potentiality. It was made evident that hydrated H⁺ ions and hydrated H⁻ ions could exist in the same aqueous solution without interaction. There was the suggestion that metabolic processes which included hydrogen commonly involved factors effecting the dehydration of hydrogen ions.

A study of the solubility of hydrogen gas in water yielded results which were in accord with behavior patterns previously indicated for nitrogen gas and oxygen gas in aqueous solution. It thus appeared that real progress had been made in the interpretation of the relationships of gases to an aqueous solvent.

CHAPTER 14.   CONCERNING CARBON AS A METABOLITE

From the very beginning of the organic world carbon has had a major role as a motherly element, nurturing through association the development of chemical compounds of seemingly infinite variety. The common primary associates of carbon have been the elements oxygen and hydrogen. In foregoing chapters atomic oxygen and atomic hydrogen were evidenced as capable of being metabolically operative in roles as positive or negative ions in hydrated or in anhydrous states. For association it seemed obvious that atomic carbon would be indicated as capable of metabolic activity in the same manner.

In chemical science atomic carbon often had been treated as though it possessed covalent potentialities when ionic. The discovery that osmosis could bring about dissociation led to a specific characterization of the components of molecules as sub-ionic atoms, either positive or negative but never both positive and negative. This characterization made imperative a designation of the electrical nature of each component atomic ion as well as of the valence in the projection of structural formulas. Such a designation in turn led to the recognition of greater symmetry in numerous formulations and presumably in nature. An outstanding example of such a development with special reference to carbon was given in the structural formula for chlorophyll as represented in a preceding chapter.

The present study of carbon as a metabolite was a natural sequel to previous studies relating to metabolic nitrogen, oxygen and hydrogen. Similarly the study was projected from a background of familiarity with the description of periodic hydrational potentiality as given and repeatedly validated in the book "Behavior Patterns of Hydration".

THE CARBONATE RADICAL

In the earth's crust carbonates have been evidenced as abundant, and commonly many have been appraised as having originated through extrusive deposition. As limestones, dolomites and corals the depositions have involved the activities of organisms : plants or animals or both plants and animals. As extrusive depositions the carbonates have been subject to characterization as end products of metabolism, intrinsically inert and inorganic. In foregoing chapters it was suggested that electrons were synonymous with energy. In keeping with this suggestion and the characterization of carbonates as end products of metabolism it appeared significant that the sub-ionic atoms of carbon in carbonates were evidenced as positively charged four-valent, and thus devoid of accessory electrons potentially subject to release an energy.

In chemistry the structural formula commonly projected for the carbonate radical CO_3^{--} has not included a full characterization of the constituent sub-ionic atoms. The two types of representation have been given in Figure 12.

At least in some associations the anionic 2-valent solute carbonate radical clearly has remained intact and has become hydrated precisely as prescribed by the description of periodic hydrational potentiality. Such a development was subject to documentation by the data of specific gravity. Illustrative data for some aqueous solutions of sodium carbonate and potassium carbonate have been given in Table 16.

As a 2-valent anion it was natural to assume that each of the three sub-ionic atoms of oxygen in the carbonate radical bore two negative charges and that the single sub-ionic atom of carbon bore four positive charges. That such an assumption was correct was evidenced by the osmotic dissociation of the radical, since under such conditions each component sub-ionic atom was released and each such released atomic element ion hydrated to the precise extent prescribed by the description of periodic hydrational potentiality.

Table 16. Specific gravity data for aqueous solutions of two carbonates.

Solute Ions	Wa	H	Wh	$\dfrac{Wa}{Wh}$
Na⁺	24	11	222	
Na⁺	24	11	222	.126
CO₃⁻⁻	56	18	380	
	104		824	

Gms. Na₂CO₃ per liter	÷.126	÷1.126	Ml Solvent per liter	Calc. Sp. Gr.	Obs. Sp. Gr.
10.09	80.0	71.0	929	1.009	1.0086
20.38	161.8	143.5	856.5	1.0183	1.0190
41.59	329.8	292	708	1.0378	1.0398
63.64	505	448	552	1.057	1.0606
110.30	876	778	222	1.098	1.1029
160.50	1273	1132	-132	1.141	1.1463

Solute Ions	Wa	H	Wh	$\dfrac{Wa}{Wh}$
K⁺	40	3	94	
K⁺	40	3	94	.2395
CO₃⁻⁻	56	18	380	
	136		568	

Gms. K₂CO₃ per liter	÷.2395	÷1.2395	Ml Solvent per liter	Calc. Sp. Gr.	Obs. Sp. Gr.
20.33	85	68.6	931.4	1.0164	1.0163
41.38	172.5	139.2	860.8	1.0333	1.0345
63.17	264	213	787	1.051	1.0529
85.72	358	289	711	1.069	1.0715
109.0	455	367	633	1.088	1.096
133.2	556.5	450	550	1.1065	1.1096

Data for the osmotic behavior of the two carbonates considered in Table 16 were included in Table 14 of the mentioned book dealing with hydration. For convenience these data in an abbreviated form have been given in Table 17.

The evidenced osmotic dissociation of the carbonate radical characterized the involved sub-ionic atom of carbon as four-valent,

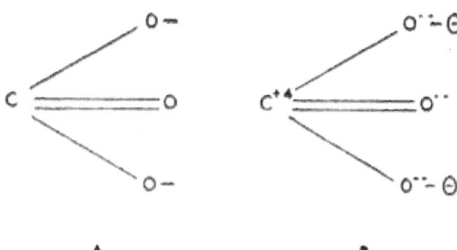

Figure 12. Two types of structural formulas projected for the carbonate radical.

A=the conventional type of chemical science.

B=the type with full characterization of the constituent sub-ionic atoms.

Table 17. Data on the osmotic behavior of some carbonates, evidencing a complete dissociation of the radicals and a complete hydration of the resultant element ions.

Solute	Assumed Ions	Wa	H	Wh	Calc. Specific Vol. Increase	Total Calc. Sp. Vol. Inc.	Observed Specific Vol. Increase
Na_2CO_3	Na^+	24	11	222	88		
	Na^+	24	11	222	88	313.5	725
	CO_3^{--}	56	18	380	137.5		
	Na^+	24	11	222	88		
	Na^+	24	11	222	88		
	C^{+4}	20	13	254	108	725	725
	O^{--}	12	17	318	147		
	O^{--}	12	17	318	147		
	O^{--}	12	17	318	147		
K_2CO_3	K^+	40	3	94	13		
	K^+	40	3	94	13	163.5	575
	CO_3^{--}	56	18	380	137.5		
	K^+	40	3	94	13		
	K^+	40	3	94	13		
	C^{+4}	20	13	254	108	575	575
	O^{--}	12	17	318	147		
	O^{--}	12	17	318	147		
	O^{--}	12	17	318	147		

positive and hydrated with 13 H_2O units as prescribed for a weight of 20 in the anhydrous state. The data carried the obvious implication that carbon C^{+4} ions in both anhydrous and hydrated states had potential roles as metabolites, since all protoplasm was directly or indirectly associated with the osmotic potentialities of vacuolar tissue.

The role of C^{+4} ions in synthetic processes was subject to appraisal as a most important one involving the areas of structure and bondage. Carbonates and carbon dioxide comprised the chief source of inorganic carbon for synthesis,—and in these compounds the sub-ionic atoms of carbon were of this nature. The data which deprecated covalent potentiality for ionic atoms made it quite clear that sub-ionic C^{+4} atoms were common constituents of organic compounds, since every carbon to carbon linkage of 4-valent carbon atoms included one.

Commonly it had been found convenient to group carbonates and carbon dioxide as sources of carbon in relation to the synthesis of carbohydrates by bacteria and by green plants. Neither the relationship of the two substances nor their relative importance was clearly defined. The presence of carbonates in sea water and the gross photosynthetic activity of algae seemed to emphasize the potential importance of solute carbonate radicals as a source of carbon, but the release of carbon dioxide in the respiration of plants and animals similarly emphasized its importance. The interrelationship of the carbonate radical and carbon dioxide thus became a matter of special interest.

CARBON DIOXIDE

In the reaction of acids on carbonate, the carbon sub-ionic atoms of the released carbon dioxide presumably would be four-valent positive, C^{+4}, as would also be the carbon in carbon dioxide released from the combustion of organic materials and from respiration. In substantial measure the carbon in carbon dioxide, as noted, appeared to be the carbon which was fundamental in the elaboration of carbohydrates by the higher green plants covering extensive land areas of the earth.

A study of the solubility of carbon dioxide in water led to evidence that the behavior pattern of the gas was somewhat exceptional in that it involved ionization without a preceding dissociation. The indicated ions appeared to hydrate in a manner precisely conforming to the description of periodic hydrational potentiality. On the assumption that the extraneous carbon dioxide gas was neutral the suggested change which took place in water was the one represented in Figure 13.

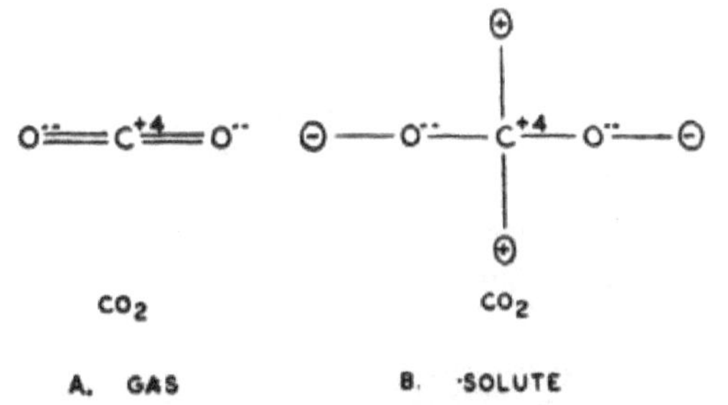

Figure 13. Suggested development with respect to carbon dioxide in water.

As the electrically neutral and thus non-electrolytic solute unit represented in Figure 13 the calculated ionic weight in the anhydrous state would be 44, and in consequence the ion would have, under the description of periodic hydrational potentiality, the ability to take on and hold 1 H_2O^- unit in hydration. It was ventured that the nature of the solute, including its minimal hydrational potentiality, accounted for its anomalous unstable behavior pattern. The calculated density of the solute would be 1.71 and a liter thus would weigh 1710 grams. The amount of carbon dioxide soluble in a liter of water commonly has been reported as 1713 milliliters. Introduced into water under pressure the gas had an apparent increase in solubility and under electrical stress the liquid became conductive. The instability of the projected CO_2^{+2-2} solute unit was supported by the calculation for its pattern of behavior in osmosis, the derived value being minus four, representing a slight reduction in solution volume.

The order of agreement between the above-indicated values 1710 and 1713 seemed to represent more than accidental and insignificant coincidence, but no immediate explanation was forthcoming. There was, however, the suggestion that in some manner the projected non-electrolytic CO_2^{+2-2} unit was involved. This suggestion prompted a search for further and more definite evidence of the existence of such a radical. The search was directed to analyses of the osmotic behavior patterns of simple carbohydrate radicals because of the surmise that the CO_2^{+2-2} unit, if present, might be subject to direct utilization in synthesis without dissociation into element ions. It seemed obvious that such a unit would be amphoteric and potentially covalent, and it was ventured that possibly the behavior pattern of such a unit had been involved in the common chemical assignment of covalency to element ions of carbon.

THE TARTRATE RADICAL

For many years the tartrate radical had been appraised as having the composition $C_4H_4O_6$. When a complete characterization of the sub-ionic atomic components of the radical was attempted, however, the correct composition seemed more likely to be $C_4H_2O_6$. Since the description of periodic hydrational potentiality had prescribed a weight of 2 for neutral hydrogen atoms and chemical science had prescribed a weight of 1 or thereabouts for hydrogen atoms irrespective of category or state, the weight of the tartrate radical was essentially the same in the two appraisals. The structural formula considered satisfactory for the tartrate radical appraised as of the composition $C_4H_2O_6$ has been given in Figure 14.

An examination of the structural formula given in Figure 14

mide it readily apparent that a disruption of the carbon-to-carbon linkages would yield solute units as follows : $CH_2O_2^{--}$, CO^{--}, CO^{--}, CO_2^{+2-2}. It was to be recognized that of these listed units the CO_2^{+2-2} unit was identical with the unit projected as formed on the introduction of carbon dioxide into water. It was to be recognized further that following a disruption of the carbon-to-carbon linkages each of the ionic units formed would be expected to hydrate. In true solutions they would be expected to hydrate to the full extent prescribed by the description of periodic hydrational potentiality.

Rather surprisingly it was found that the specific gravity of aqueous solutions of tartrates evidenced a disruption of the carbon-to-carbon linkages within the tartrate radicals. Data for the specific gravity of some aqueous solutions of tartrates were available in widely published Tables. Representative data have been given in Tables 18 and 19

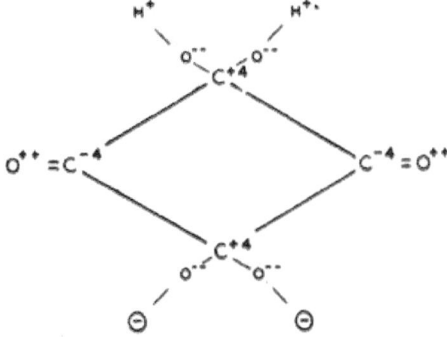

Figure 14. Structural formula projected for the tartrate radical appraised as of the composition $C_4H_2O_6$.

It was clear from the data of Table 18 and 19 that in aqueous solution the carbon-to-carbon linkages of the projected tartrate radical had been broken and that the resultant ions had become hydrated precisely in the manner prescribed for them by the description of periodic hydrational potentiality. There was also supplementary evidence of the persistence of the involved hydrational attraction in the absence of free solvent.

Table 18. Data pertaining to the specific gravity of sodium tartrate in aqueous solution.

Assumed Ions	Wa	H	Wh	Solute : $Na_2C_4H_2O_6$.	4 H_2O
Na^+	24	11	222		
Na^+	24	11	222	$\dfrac{192}{192+(4\times18)}$	$=.727$
$CH_2O_2^{++}$	52	20	412		
CO^{--}	24	11	222		
CO^{--}	24	11	222	$\dfrac{192}{1362}=.141$	
Co_2^{+2-2}	44	1	62		

Gms. Solute per liter	$\times.727$	$+.141$	$+1.141$	Ml Solvent	Specific Gravity Calc.	Obs.
62.46	45.4	322	282	718	1.040	1.0480
107.0	77.8	552	484	516	1.068	1.0702
178.5	130	922	808	162	1.114	1.1156
229.4	167	1182	1035	- 35	1.147	1.1471
283.1	206	1460	1280	-280	1.180	1.1797
339.7	247	1752	1535	-535	1.217	1.2132

Table 19. Data pertaining to the specific gravity of sodium potassium tartrate in aqueous solution.

Assumed Ions	Wa	H	Wh	Solute : $NaKC_4H_2O_6$.	H_2O
Na^+	24	11	222		
K^+	40	3	94	$\dfrac{208}{208+(4\times18)}=742$	
$H_2CO_2^+$	52	20	412		
CO^{--}	24	11	222		
CO^{--}	24	11	222	$\dfrac{208}{1234}=.1685$	
CO_2^{+2-2}	44	1	62		
	208		1234		

Gms. per liter	$\times.742$	$+.1685$	$+1.1685$	Ml Solvent	Specific Gravity Calc.	Obs.
62.34	46.3	275	236	764	1.039	1.0390
84.24	62.6	374	320	680	1.054	1.0530
106.70	79.0	469	402	598	1.067	1.0673
129.80	96.3	572	490	510	1.082	1.0818
153.50	114.0	677	580	420	1.097	1.0965
177.80	132.0	784	672	328	1.112	1.1114

Included among the ions released on the disruption of the carbon-to-carbon linkages was the amphoteric non-electrolytic ion which had been projected as formed when carbon dioxide was in contact with water. In the present instance, however, this ion was evidenced as stable, in contrast to the instability exhibited when alone. There thus was the suggestion and evidence that when accompanied by certain other ions the CO_2^{+2-2} unit became a stable entity. It seemed to follow that under conditions commonly prevailing in protoplasm the unit would have a varying yet appreciable degree of stability and potential usefulness as a metabolite.

It was recognized that within protoplasm the radicals containing carbon which had been projected and evidenced in Tables 18 and 19 might be present, or might have been modified by osmotic membranes. It became of interest, therefore, to examine the osmotic behavior pattern of sodium potassium tartrate. The results obtained in such a study were assembled to comprise Table 20.

Table 20. Data relating to the osmotic behavior of sodium potassium tartrate.

Assumed Ions	Wa	H	Wh	V_1	Total V_1	Observed V_1
Na^+	24	11	222	88		
K^+	40	3	94	13		
$H_2CO_3^{++}$	52	20	412	153.5	426.5	630
CO^{--}	24	11	222	88		
CO^{--}	24	11	222	88		
CO_2^{+2-2}	44	1	62	(-4)		
Na^+	24	11	222	88		
K^+	40	3	94	13		
H^+	4	21	382	180.5		
H^+	4	21	382	180.5	630	630
CO_2^{+2-2}	44	1	62	(-4)		
CO^{--}	24	11	222	88		
CO^{--}	24	11	222	88		
CO_2^{+2-2}	44	1	62	(-4)		

In the upper portion of Table 20 it was to be noted that when it was assumed that all of the radicals remained intact in osmosis there was no agreement between the calculated and observed values for volumetric increase.

In the lower portion of Table 20 it was to be noted that when it was assumed that osmotic dissociation potentialities had effected the removal of two sub-ionic atoms of hydrogen, which thereupon had become hydrated, there was a precise agreement between the calculated and observed values for volumetric increase. The indicated activity resulted in the formation of additional CO_2^{+2-2} units identical with the units projected as formed on the contact of CO_2 with water.

Collectively the data of Tables 18, 19 and 20 carried the distinct implication that protoplasmic metabolites included simple associations of carbon and oxygen as hydrated ions whose electrical charges might be balanced—the non-electrolytic status—or might be unbalanced—the electrolytic status. It was considered of interest and potential significance that one of these associations, CO_2^{+2-2}, was identical with the solute unit suggested or evidenced as having been formed on the contact of carbon dioxide and water. It was considered of interest further that the sub-ionic C^{-4} atoms indicated as constituents of the tartrate radical represented the capture and retention of eight electrons by C^{+4} ions.

THE ACETATE RADICAL

As noted previously, chemical science had characterized ionic atomic carbon as having two common forms : 2-valent and 4-valent. It was obvious that up to this point the considerations had involved 4-valent carbon, and it became of interest to examine the behavior pattern of 2-valent carbon. Sub-ionic 2-valent carbon appeared to be included in the acetate radical on the basis of the evidence to follow.

In conventional chemistry the compositional and structural formulas for the acetate radical were those indicated herewith in Figure 15.

Composition : $C_2H_3O_2$

Structure :

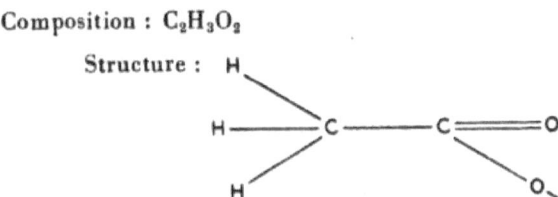

Figure 15. Conventional compositional and structural formulas for the acetate radical.

The structural formula given in Figure 15 did not include a full characterization of the component sub-ionic atoms, and when such a characterization was ventured it became obvious that the formula was unacceptable. Each of the three sub-ionic atoms of hydrogen was subject to replacement by F⁻, Cl⁻, Br⁻, or I⁻ ions, on which account it was concluded that these hydrogen atoms bore negative charges. From this position it followed that the adjacent sub-ionic atom of carbon was 4-valent positive. The next adjoining sub-ionic atom of carbon in the formula as given would be 4-valent negative and the adjoining oxygen atoms would be 2-valent positive. This relationship would give the radical a positive charge contrary to observed behavior.

Before venturing to derive a more acceptable structural formula for the acetate radical it was of interest to study its behavior in aqueous solution. Specific gravity offered the simplest approach in such a study,—and observational data on the specific gravity of aqueous solutions of potassium acetate were available in widely published tables. Excerpts from one of these tables have been correlated with the description of periodic hydrational potentiality in Table 21.

From the data given in Table 21 it was concluded that the involved compositional formula for the acetate radical was correct and that the radical had remained as an intact stable unit in the indicated aqueous solutions. This behavior pattern was in contrast to the behavior pattern evidenced for the tartrate radical in which a disruption of the carbon to carbon linkages had been indicated.

Table 21. Data pertaining to the specific gravity of potassium acetate in aqueous solution.

Assumed Ions	Wa	H	Wh	Solute : $KC_2H_3O_2 + 4\ H_2O$
K	40	3	94	$\dfrac{100}{100+(4\times 18)} = .582$
$C_2H_3O_2$	60	16	348	
	$\overline{100}$		$\overline{442}$	$\dfrac{100}{442} = .226$

Gms. Solute per liter	$\times .582$	$\div .226$	$\div 1.226$	Ml Solvent	Specific Gravity Calc.	Obs.
40.764	23.7	105	85.8	914.2	1.0192	1.0191
61.758	35.9	159	130	870	1.0290	1.0293
83.16	48.5	214.5	175	825	1.0395	1.0395
104.97	61	270	220	780	1.0500	1.0497
127.188	74	327.5	267	733	1.0605	1.0599
149.842	87.2	386	314.8	685.2	1 0712	1.0703

Since the conventional structural formula for the acetate radical had been found to be incompatible with a full characterization of the constituent sub-ionic atoms it seemed imperative to study the osmotic behavior. For this purpose an aqueous solution of potassium acetate was used. A specific molecular osmotic volumetric increase of 337 milliliters was obtained. The calculated volumetric increase for the solute with an assumed K^+ ion and an assumed intact $C_2H_3O_2^-$ ion was 131.5 milliliters. It was obvious that osmosis had brought about some dissociation of the acetate radical. The nature of this dissociation was interpreted to be that indicated in the data of Table 22.

From the data given in Table 22 it was possible to construct a structural formula for the acetate radical which would on the one hand permit a satisfactory complete characterization of the constituent sub-ionic atoms and on the other hand permit an identification of the linkage which were broken in osmosis. These developments have been indicated in Figure 16.

Table 22. Data pertaining to the osmotic behavior of the acetate radical in aqueous solutions of potassium acetate.

Assumed Ions	Wa	H	Wh	V_1	Total V_1	Observed V_1
CH_3^+	20	13	254	108		
CO^{-+}	28	9	190	69		
O^{--}	12	17	318	147	337	337
K^+	40	3	94	13		

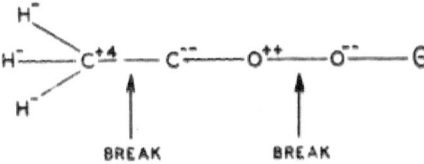

Figure 16. Structural formula for the acetate radical projected from the observed pattern of osmotic behavior.

In conjunction with the developments represented in Figure 16 it was of interest that the CH_3 moiety appeared to remain intact in osmosis, since the NH_4 radical, similarly with sub-ionic H atoms, had remained intact under similar circumstances. It was of interest further than an O^{--} ion had been evidenced as released, since such ions had been indicated as released in the osmotic dissociation of such radicals as carbonates, sulfates and phosphates. It was of interest still further that a non-electrolytic covalent CO^{+-} ion had been evidenced as released, since such an ion had obvious analogy with the CO_2^{+2-2} ion considered previously. The projected sub-ionic C^{--} atom in the acetate radical represented the capture and retention of six electrons by a metabolic C^{+4} ion.

DISCUSSION

As made evident in the foregoing data, the combined services of specific gravity and osmosis as research tools proved to be of substantial value in relation to a study of the prospective behavior of carbon as a metabolite. The data were interpreted as evidencing for the C^{+4} ion, potentially hydrated with 13 H_2O units as prescribed by the description of periodic hydrational potentiality, a basic role within the protoplasmic matrix. It could remain in this 4-valent positive ionic state subject to interpretation as inorganic and devoid of accessory electrons expendable as energy, or it could acquire electrons to attain in succession the ionic states C^{+2}, C^{-2} and C^{-4}. Its acquisition of electrons appeared to depend upon its association with excited ions of oxygen and hydrogen, and in these associations its role seemed primarily that of a storage reservoir mediated through bondage both chemical and hydrational. As thus appraised, metabolic carbon seemed to invite supplementary speculation.

Previously it was pointed out that from the repeatedly validated description of periodic hydrational potentiality the prescribed attributes of H^- ions conferred for the anhydrous state an analogy with radiation by virtue of zero weight and infinite mobility. When anhydrous H^- ions were projected as constituents of solar radiation it followed that such events as friction and impact might effect the loss of the accessory expendable electron and an acquisition of weight. It followed further that specific types of interception might effect electron release under conditions making possible their immediate capture. Resonant oxygen in the nuclei of chlorophyll molecules was projected as the major type of such an interception.

Subject to correlation with such a projection was the evidence of a sub-ionic 8-valent negative oxygen atom in biatomic atmospheric oxygen, assumed to be identical with the biatomic oxygen released in photosynthesis. In conjunction with the study of osmosis it was noted that this process involved the randomized exit of hydrated ions from all adjoining membrane-liquid interfaces. Under such conditions the internal osmotically mediated increase in solution volume was brought about by one-half the solute originally present. Since photosynthesis took place within a matrix which included membranes it was of interest to note in passing that an external release of biatomic oxygen containing a sub-ionic electron-rich O^{-8} atom might also be accompanied by an internal release of identical material serving as a source of electrons to facilitate synthetic reactions. The considerations obviously led back to the still-challenging mysteries of photosynthesis.

When the foregoing considerations were correlated with the behavior patterns of light-sensitive chlorophyll-containing seeds there followed the suggestion that radiation at 6867.2 Angstroms excited oxygen atoms in chlorophyll in such a manner as to effect both the acquisition and the release of electrons, whereas radiation at 7593.8 Angstroms had a counteracting effect. As thus projected oxygen was the element primarily involved in photosynthesis through its capture, retention and release of electrons brought by negatively-charged radiation hydrogen. Yet ionic carbon was an essential ally for the capture and release or indefinite retention through association of the electrons released by excited oxygen.

A supplementary product of this study was an enhanced appreciation of the potential value of specific gravity and osmosis as tools of research. Heretofore these attributes had served to document the validity of the description of periodic hydrational potentiality. With an established validity for the description as a background they served to reveal otherwise obscure developments within aqueous solutions.

SUMMARY

Using specific gravity and osmosis as research tools in conjunction with a description of periodic hydrational potentiality some behavior patterns of solute carbon atoms and simple carbon-containing ions were examined. The results obtained suggested that metabolically organic synthesis began with C^{+4} ions devoid of accessory expendable electrons and proceded through the acquisition of electrons to involve as progressive stages the ions C^{++}, C^{--} and C^{-4}. The C^{+4} ions were demonstrable as hydrated and stable in aqueous solution following the osmotic dissociation of solute carbonate radicals. The C^{+4} ions were demonstrable also as sub-ionic units of solute ionic carbon dioxide. The specific gravity and osmotic behavior patterns of some aqueous solutions evidenced the formation of a variety of solute units in which carbon was associated with oxygen or hydrogen or both oxygen and hydrogen. The acquisition of accessory expendable electrons by C^{+4} ions was appraised as the major role of metabolic carbon.

CHAPTER 15.   CONCERNING SULFUR AND PHOSPHORUS AS METABOLITES

1. SULPHUR

The description of periodic hydrational potentiality, given and repeatedly validated in a preceding monograph, provided hitherto unavailable avenues of approach to numerous problems involving behavior within an aqueous matrix. These approaches were used in previous chapters concerned with the respective behavior patterns of nitrogen, oxygen, hydrogen and carbon, and supplied the procedures followed in the studies herewith reported.

Sulfur has had an unusually interesting relationship to civilization, among other things having been a prominent feature of an imagined hell awaiting the wicked. In chemical science sulfur has been a noteworthy element in several ways including its numerous valence forms, its diverse S_8 molecules, its effectiveness as a catalyst and its relations to light, heat and acidity. As a metabolite sulfur commonly has appeared to have at least two roles: it has been identified through chemical analyses as a constituent of some organic compounds, and in combination with hydrogen it has been released from organic media incident to anaerobic respiration and has thus been revealed as a potential metabolic substitute for oxygen. The present study was concerned with the behavior patterns of sulfur within an aqueous matrix.

The description of periodic hydrational potentiality prescribed that an S^{+6}ion would have a weight of 44 and an ability to take on and hold $1H_2O^-$ unit in hydration. These prescribed attributes were found to be quite readily subject to demonstration as valid. In a simple aqueous solution of copper sulfate the hydrated S^{+6} ion was evidenced as present in a stable state. Moreover, at concentrations of copper sulfate in which no free solvent was present the hydrated ions were evidenced as maintaining individuality while sharing the involved H_2O^- units. Data relating to these respective developments have been published in earlier papers, but for convenience have been repeated herewith in Table 23.

Table 23. Data pertaining to the behavior of copper sulfate in water. Upper section; basic data. Center section; free solvent present, Lower section; all solvent hydrationally bound,

Assumed Ions	Wa	H	WH	
Cu^{++}	62	15	332	
S^{+6}	44	1	62	$\dfrac{154}{1666} = .0925$
O^-	12	17	318	
O^{--}	12	17	318	
O^{--}	12	17	318	
O^{--}	12	17	318	
	$\overline{154}$		$\overline{1666}$	

Gms/liter	÷.0925	÷1.0925	Ml Solvent	Calc. Sp. Gr.	Obs. Sp. Gr,
26.22	283.5	259.3	740.7	1.0242	1.0254
46.76	527	482	518	1.045	1.0450
60.88	658	602	398	1.056	1.0582
75.35	815	746	254	1.069	1.0716
90.20	976	894	106	1.082	1.0854
97.76	1057	966	34	1.091	1.0923
113.2	1224	1120	−120	1.104	1.1063
129.0	1394	1275	−275	1.119	1.1208
145.2	1570	1437	−437	1.133	1.1354
178.9	1933	1770	−770	1.163	1.1659
196.4	2125	1944	−944	1.181	1.1817
214.0	2319	2115	−1115	1.204	1.206

The data given in Table 23 were presented as evidence that the hydrated S^{+6} ion could exist as a stable entity in an aqueous solution. However, its occurrence seemed predicated on the presence of Cu^{++} ions, since in the case of most water soluble sulfates the data of specific gravity clearly have evidenced the presence of intact sulfate SO_4^{--} radicals hydrated with 23 H_2O^- units as prescribed for an ion of weight 92 by the description of periodic hydrational potentiality. In view of this relationship there was the suggestion that the copper Cu^{++} ion might dissociate intact sulfate radicals and thus release atomic S^{+6} ions not otherwise available. With respect to metabolism, however, the suggestion was of little significance, since osmosis brought about the complete dissociation of all solute sulfate radicals, as was made evident in the data of Tables 14 and 15 of the book "Behavior Patterns of Hydration." All metabolically active protoplasm was appraised as having a direct or indirect association with vacuolar tissue, on which account it followed that the common pattern of entrance for sulfate sulfur into the metabolic matrix was that of the S^{+6} ion accompanied by $1H_2O^-$ unit in hydration. As thus interpreted the S^{+6} ion became somewhat analogous to the C^{+4} ion considered in a previous chapter as devoid of accessory readily expendable electrons subject to release as energy and as thus within the category of the inorganic.

As noted previously, when specific gravity was used as an index of solute status the sulfate radicals in aqueous solutions of sodium salts became indicated as having remained intact. Further attention was directed to these sulfur containing salts which represented sulfur as the sub-ionic atoms S^{+6} and S^{+4}. The S^{+4} ion was subject to interpretation as a potential derivative of the S^{+6} ion through the accession of a pair of electrons. This attention revealed that in the indicated solutions the radicals not only remained intact but apparently were involved in the development of an interesting situation in which there was no sharing of the H_2O^- units held by the ions in hydration. In the absence of sharing it followed that solubility was conditioned and limited by the collective hydrational potentialities of the involved ions, -or conversely, that the limits of solubility attested the absence of sharing. Because of the indicated relationship it was possible by extrapolation to calculate data for saturated solutions. Data illustrative of the indicated behavior patterns have been given in

Tables 24 and 25.

Table 24. Data pertaining to the behavior of sodium sulfate in water. Upper section; basic data. Lower section: free solvent present to saturation.

Solute	Assumed Ions	Wa	H	Wh	
$Na_2SO_4^-$	Na^+	24	11	222	$\frac{140}{950}=.147$
	Na^+	24	11	222	
	SO_4^{--}	92	23	506	
		140		950	

Gms/liter	$\div.147$	$\div 1.147$	Ml Solvent	Calc. Sp. Gr.	Obs. Sp. Gr.
50.79	345	301	699	1.044	1.044
70.32	478	418	582	1.060	1.060
90.50	616	538	462	1.078	1.077
116.5	792	692	308	1.100	1.098
149.3	1015	886	114	1.129	1.125
168.5	1147	1000	0	1.147	

Table 25. Data pertaining to the behavior of sodium sulfite in water. Upper section ; basic data. Lower section, free solvent present to saturation.

Solute	Assumed Ions	Wa	H	Wh	
Na^2SO_3	Na^+	24	11	222	$\frac{124}{664}=.1865$
	Na^+	24	11	222	
	SO_3^{--}	76	8	220	
		124		664	

Gms./liter	$\div.1865$	$\div 1.1865$	Ml Solvent	Calc. Sp. Gr.	Obs. Sp. Gr.
20.34	109	92	908	1.017	1.0172
133.8	718	606	394	1.112	1.1146
158.8	850	716	284	1.134	1.1346
184.8	990	834	166	1.156	1.1549
211.6	1135	956	44	1.179	1.1755
221.0	1186.5	1000	0	1.1865	

From the data of Tables 24 and 25 it was concluded that when associated with sodium the sulfate and sulfite radicals remained intact in aqueous solution and hydrated to the precise extent prescribed by the description of periodic hydrational potentiality. In conjunction with the data of Table 23 there was the suggestion that the radicals would be subject to dissociation by Cu^{++} ions, but insofar as metabolism was concerned both the involved S^{+6} ions and S^{+4} ions, as noted, had been evidenced as released in osmosis and hence made available as metabolites.

Collectively the data of Tables 23, 24, and 25 indicated that the presence or absence of an ability on the part of solute ions to share the H_2O^- units held in hydration was a challenging subject for further study since it contained prospects for an improved understanding of solubility. It was as obvious, however, that at this point such a study would be a digression.

It seemed clear that the acquisition of two electrons by an S^{+4} ion would yield an S^{+2} ion and attention was directed to the behavior patterns of such S^{+2} ions, or of S^{+2} ions irrespective of the nature of their derivation. These ions as sub-ionic atoms were assumed to be present in the thiosulfate radical $S_2O_3^{--}$. Data pertaining to the specific gravity of aqueous solutions of sodium thiosulfate have been given in Table 26.

From the data of Table 26 it was concluded that in aqueous solution the chemical bondages of the thiosulfate radical were broken to effect a complete dissociation, and that immediately following this dissociation the released element ions hydrated to the precise extent prescribed by the description of periodic hydrational potentiality. The included S^{+2} ions were evidenced as having maintained individual and characteristic behavior patterns, as was the case with their indicated associates.

Up to this point the sulfur ions considered have been positively charged—S^{+6}, S^{+4} and S^{+2}. It seemed obvious that the acquisition of four electrons by an S^{+2} ion would yield an S^{-2} ion. Bivalent

Table 26. Data pertaining to the specific gravity of aqueous solutions of sodium thiosulfate. Upper portion; basic data, including allowance for 2 H_2O^- units in original salt. Central portion; free solvent present. Lower portion; hydrational bondage

Assumed Ions	Wa	H	Wh	
Na^+	24	11	222	$\dfrac{156}{156+(2\times18)} = .813$
Na^+	24	11	222	
SS^{+2}	36	5	126	
S^{+2}	36	5	126	
O^{--}	12	17	318	$\dfrac{156}{1650} = .0946$
O^{--}	12	17	318	
O^{--}	12	17	318	
	156		1650	

Gms./liter	×.813	÷.0946	÷1.0946	Ml Solvent	Calc. Sq. Gr.	Obs. Sq. Gr.
20.30	16.5	174	159	841	1.015	1.0148
39.44	32.0	338	309	691	1.029	1.0317
60.07	48.8	516	472	528	1.044	1.0476
74.17	60.2	636	582	418	1.054	1.0584
95.89	78.0	824	754	246	1.070	1.0751
110.70	90.0	958	870	130	1.080	1.0863
132.0	107.5	1135	1087	-37	1.098	1.1003
156.5	127.3	1347	1230	-230	1.117	1.1182
181.8	148	1563	1430	-430	1.133	1.1365
207.9	169	1787	1630	-630	1.157	1.1551
234.8	191	2020	1845	-845	1.175	1.1740
262.5	213	2250	2055	-1055	1.195	1.1932

negative sulfur ions were assumed to be present in sulfides as sub-ionic atoms. Data pertaining to the specific gravity of aqueous solutions of sodium sulfide have been given in Table 27.

Table 27. Data pertaining to the specific gravity of aqueous solutions of sodium sulfide. Upper portion; basic data. Central portion; free solvent present. Lower portion; hydrational bondage.

Assumed Ions	Wa	H	Wh	
Na⁺	24	11	222	$\frac{76}{634} = .12$
Na⁺	24	11	222	
S⁻⁻	28	9	190	
	76		634	

Gms./liter	÷ .12	÷ 1.12	Ml Solvent	Calc. Sq. Gr.	Obs. Sq. Gr.
10.10	84.1	75.2	924.8	1.0088	1.0098
20.42	170	152	848	1.218	1.0211
41.76	348	311	689	1.037	1.044
64.03	534	477	523	1.057	1.0672
87.26	726	648	352	1.078	1.0907
111.5	928	830	170	1.098	1.1146
136.7	1137	1013	-13	1.124	1.1388
162.9	1357	1210	-210	1.127	1.1634
190.2	1583	1413	-413	1.170	1.1885
218.5	1820	1622	-623	1.197	1.214

An examination of the data of Table 27 revealed that although the order of agreement between calculated and observed specific gravity values was not as satisfactory as that represented in the data of the four preceding tables there nevertheless was rather clear evidence that the involved ions were hydrated and that hydrational bondages persisted in the absence of free solvent. As a further check on the integrity of the solute hydrated S⁻⁻ion it became of interest to test the osmotic behavior of sodium sulfide. The results obtained have been given in Table 28.

From the data given in Table 28 it was clear that the S⁻⁻ ion was present in aqueous solutions of sodium sulfide and hydrated to the precise extent prescribed by the description of periodic hydrational potentiality. The release of hydrogen sulfide from organic matter carried the indirect implication that the S⁻⁻ ion was a potential metabolite, but supplied no indication of its behavior pattern in hydration.

Table 28. Results obtained in a study of the osmotic behavior of sodium sulfide.

Solute	Assumed Ions	Wa	H	Wh	Vh	Va	D_1	$\dfrac{D_1}{2}$	$\dfrac{Va}{2}$	D_2	Calc. Standard Volumetric Increase	Observed Standard Volumetric Increase
Na₂S	Na⁺	94	11	222	200	12	188	94	6	88	245	245
	Na⁺	24	11	222	200	12	188	94	6	88		
	S⁻⁻	28	9	190	166	14	152	76	7	69		

Legend : Column headings as in "Behavior Patterns of Hydration".

Some years ago in conjunction with litigation over alleged crop damage by smelter fumes it became obvious that in the vicinity of numerous industrial plants the sulfur content of soils was increased. A substantial portion of this increase was attributed to the interaction of sulfur dioxide and water. It became of interest, therefore, in relation to an appraisal of metabolic sulfur, to examine the solubility of sulfur dioxide in water from an hydrational background.

For the indicated examination there were three major assets: the description of periodic hydrational potentiality, a previously-reported study of the solubility of carbon dioxide in water, and published data on the amount of sulfur dioxide soluble in a liter of water. These assets were useless without speculative probing, which always introduced personal interpreting. One way of accunting for the solubility of sulfur dioxide in water was as follows.

It was assumed that sulfur dioxide was an electrically neutral gas subject to representation with a structural formula

$$O^{--} = S^{+4} = O^{--}.$$

It was assumed that in water the gas became ionic and subject to representation with a structural formula

$$- -O^{--} - \overset{+}{S^{+4}} - O^{--} - -$$
$$+$$

As the indicated non-polar ion the weight in the anhydrous state would be 64. It would take on 14 H_2O^- units in hydration. In its hydrated state the weight would would be 316.

The volume of the hydrated ion would be calculable with the formula $d = 1 + \dfrac{Wa}{Wh}$. The volume would be 263 ml.

One gram molecule of dissolved sulfur dioxide would occupy a volume of 263 ml.

A liter volume would have a calculated capacity of $1000 \div 263$, or 3.8 gram molecules.

Under standard conditions 3.8 gram molecules *as a gas* would occupy $3.8 \times 22,400$ ml, or 85,120 ml.

The presence of solvent was interpreted as having prevented the attainment of the full calculated value of 85,120 ml of soluble sulfur dioxide. The minimal solvent was projected as 1 $H_2O°$ unit. The following calculations were of interest.

$$\frac{20}{316} = .063 \qquad .063 \times 85,120 = 5362.56$$

$$85,120 - 5362.56 = 79,757.44$$

The solubility of sulfur dioxide in water at °0C and standard pressure has been reported in various tables as 79,789 ml per liter. The order of agreement was interpreted as evidence that sulfur dioxide became a non-polar ion in water and hydrated to the precise extent prescribed by the description of periodic hydrational potentiality.

DISCUSSION

The description of periodic hydration potentiality given and extensively validated in the book "Behavior Patterns of Hydration" made available for the first time a mathematical basis for the study of solution phenomena involving an aqueous solvent. Through correlations with this description the nature of some specific sulfur ions capable of appreciable stability as solutes in an aqueous medium were identified from observational data on specific gravity and osmotic behavior. These ions were S^{+4}, S^{+4}, S^{+2} and $^-S^2$. S^{-2}

It was obvious that the stresses involved in osmosis were capable of releasing these ions from indicated sub-ionic states and that sulfur was an element conditioned by moderate stresses to a gain or loss of electrons in pairs. In general these considerations suggested that the major metabolic role of sulfur was that of a catalyst, yet it was recognized that the element had potentialities involving chemical bondage both in its own right and as a substitute for its chemical analog oxygen.

A supplementary study of the solubility of sulfur dioxide in water gave results interpreted as evidence that non-polar ions were formed and that these ions immediately became hydrated. From the observed osmotic behavior of other non-polar hydrated ions it was to be ventured that osmotic stresses would release S^{+4} ions from these solutes and that following hydration these S^{+4} ions would serve as potential metabolities.

SUMMARY

Behaviour patterns for some specific sulfur ions within an aqueous matrix were revealed by data pertaining to the specific gravity of aqueous solutions, osmotic activity and solubility. The noteworthy versatility of solute ionic sulfur, listed herewith as potentially occurring in the states S^{+6}, S^{+4}, S^{+2} and S^{-2}, was interpreted as evidencing a propensity for the gain or loss of electron pairs and a consequent major metabolic role as a catalyst.

2. PHOSPHORUS

Although the period marking the recognition of phosphorus as a nutrient element in plant and animal metabolism has appeared to be obscure it was quite evident that the introduction of the N-P-K era of chemical fertilizers into agriculture emphasized the importance of the element in plant nutrition. Indirectly and to a lesser extent there was an accompanying emphasis on the importance of phosphorus in animal nutrition. These developments naturally directed attention to the roles of phosphorus in the physiology of plants and animals. As to the specific behavior patterns of phosphorus in protoplasm, however, researchers have been handicapped by the absence of satisfying information regarding the relationships of solutes to an aqueous solvent. The recent publication of a description of periodic hydrational potentiality supplied a mathematical basis for the study of such relationships and it became of interest to use the approach of an examination of some behavior patterns of phosphorus.

As represented in the book "Behavior Patterns of Hydration" solute ions were subject to characterization with respect to their solvent in two ways: (1) through the specifie gravity of aqueous solutions and (2) through osmotically-mediated increases in solution volume. From the standpoint of metabolism both of these methods were indicated as desirable : the specific gravity measurements supplied information on the nature of the solutes in an external aqueous environment unmodified by the activities of organisms and the osmotic behavior supplied information as to the nature of the solutes within organisms having vacuolar tissue.

In general the data relating to the specific gravity of aqueous solutions as recorded in various published tables were obtained many years ago, for the most part by European investigators accorded appreciable renown for meticulous attention to detail. These data, obtained without any knowledge of prescribed goals, have merited further renown for their close approaches to the goals prescribed in conjunction with the mentioned description of periodic hydrational potentiality.

With respect to the general behavior of solutes in an aqueous solvent the developments in relation to specific gravity indicated that ionization was a basic feature of solubility, even for non-electrolytes, that dissociation could be partial or complete, and that at least one of the types of ions released in the dissociation became hydrated. For example, with respect to sulfates in aqueous solution there commonly was indicated a partial dissociation into cations and sulfate SO_4^{--} radicals, all units then hydrating immediately to the extent conditioned by their respective ionic weights. Yet in the case of copper sulfate there was evidenced a complete dissociation, including the breakdown of the sulfate SO_4^{--} radicals, and the immediate hydration of the resultant element ions. Insofar as permitted by the available solvent each released ion hydrated to the precise extent prescribed by the description of periodic hydrational potentiality. At solute concentrations so great as to preclude adequate solvent for individual freedom there commonly was evidenced a sharing of the hydrationally-bound H_2O^- increments present. Such developments supplied a valuable background to facilitate the study of the behavior patterns of phosphorus-containing solutes. It was possible to calculate specific gravity values from a grams-per-liter position independent of any direct consideration of atomic or molecular weights. The calculations could be based on each of a number of premises regarding the ions present in the solution, and commonly on only one premise would there result a satisfactory order of agreement between calculated and observed specific gravity values.

In contrast to the situation involving sulfur compounds in aqueous solution specific gravity data for phosphorus compounds in aqueous solution were found to be rather infrequent in published tables. The International Critical Tables, however, proved to contain somewhat fragmentary yet nevertheless reasonably satisfactory data in the desired category. This source supplied all of the observed values for the comparisons made possible in the following tables relating to the specific gravity of aqueous solutions.

In aqueous solutions of three phosphorus-containing compounds the observed specific gravity data evidenced a complete dissociation of the solute. The involved considerations have been given in Tables 29, 30 and 31. At concentrations of solute permitting the presence of free solvent (Table 31 and a portion of Table 29) a complete hydration of the released atomic element ions was evidenced. At concentrations of solute not permitting the presence of free solvent (Table 30 and a portion of Table 29) there was evidenced a sharing of the hydrationally bound H_2O^-. units.

In Tables 29, 30, and 31 the cited observational data were the only values reported for the indicated solutions in the mentioned source.

When the studies were extended to include the attributes of

Table 29. Data pertaining to the specific gravity of aqueous solutions of Na_3PO_4. Upper portion : basic data. Lower portion : application.

Solute	Assumed Ions	Wa	H	Wh	Ratio : $\dfrac{Wa}{Wh}$
	Na^+	24	11	222	
	Na^+	24	11	222	
	Na^+	24	11	222	
Na_3PO_4	P^{+5}	40	3	94	$\dfrac{160}{2032} = .0787$
	O^{--}	12	17	318	
	O^{--}	12	17	318	
	O^{--}	12	17	318	
	O^{--}	12	17	318	
Totals		160		2032	

Gms./Liter	÷.0787	÷1.0787	Ml Solvent	Calc. Sp. Gr.	Obs. Sp. Gr.
10.092	128	188.7	881.3	1.0093	1.0092
20.388	258.5	239.7	760.3	1.0188	1.0194
46.620	528	490	510	1.038	1.0405
63.744	808	758	242	1.050	1.0624
86.800	1100	1020	−20	1.080	1.0850
110.830	1410	1300	−300	1.110	1.1083

Table 30. Data pertaining to the specific gravity of an aqueous solution of K_3PO_4. Upper portion : basic data. Lower portion : application.

Solute	Assumed Ions	Wa	H	Wh	Ratio : $\dfrac{Wa}{Wh}$
	K^+	40	3	94	
	K^+	40	3	94	
	K^+	40	3	94	
K_3PO_4	P^{+5}	40	3	94	$\dfrac{208}{1648} = .1262$
	O^{--}	12	17	318	
	O^{--}	12	17	318	
	O^{--}	12	17	318	
	O^{--}	12	17	318	
Totals		208		1648	

Gms./Liter	÷.1262	÷1,1262	Ml Solvent	Calc. Sp. Gr.	Obs. Sp. Gr.
212.25	1680	1490	−490	1.190	1.1805

Table 31. Data pertaining to the specific gravity of aqueous solutions of $Na_4P_2O_7$. Upper portion: basic data. Lower portion: application

Solute	Assumed Ions	Wa	H	Wh	Ratio: $\dfrac{Wa}{Wh}$
$Na_4P_2O_7$	Na^+	24	11	222	
	Na^+	24	11	222	
	Na^+	24	11	222	
	Na^+	24	11	222	
	P^{+5}	40	3	94	$\dfrac{260}{3302} = .0788$
	P^{+5}	40	3	94	
	O^{--}	12	17	318	
	O^{--}	12	17	318	
	O^{--}	12	17	318	
	O^{--}	12	17	318	
	O^{--}	12	17	318	
	O^{--}	12	17	318	
	O^{--}	12	17	318	
Totals		260		3802	

Gms./Liter	$\div .0788$	$\div 1.0788$	Ml Solvent	Calc. Sp. Gr.	Obs. Sp. Gr.
10.092	127.8	118.6	881.4	1.0092	1.0092
20.380	258	239	761	1.019	1.0190
30.849	391	362.5	637.5	1.0285	1.0283
41.476	526	488	512	1.038	1.0369

aqueous solutions of substances which contained hydrogen as an associate of phosphorus it was found that the hydrogen-phosphorus-oxygen associations did not dissociate but persisted as radicals and with adequate solvent became hydrated to the precise extent prescribed for their respective ionic weights. It was found also that some of the salts had entered the solvent with the minimal one H_2O^- unit of hydration which had been evidenced previously for phosphoric acid. This was not a surprising development, though it emphasized an accessory significance for specific gravity measurements. The important item with respect to metabolic phosphorus was the relation of the presence of hydrogen to the maintenance of intact solute phosphorus-containing radicals. Data documenting these developments have been given in Table 32.

Table 32. Data pertaining to the specific gravity of aqueous solutions of some compounds containing hydrogen associated with phosphorus and oxygen.

Solute	Indicated State	Assumed Ions	Wa	H	Wh	Gms/L	Calc. Sp. Gr.	Obs. Sp. Gr.
K_2HPO_3	anhydrous	K^+	40	3	94	40.819	1.026	1.0308
		K^+	40	3	94			
		HPO_3^{--}	76	8	220			
KH_2PO_3	1 H_2O	K^+	40	3	94	40.95	1.246	1.0238
		$H_2PO_3^-$	80	6	188			
K_2HPO_4	1 H_2O	K^+	40	3	94	348.42	1.255	1.2633
		K^+	40	3	94			
		HPO_4^{--}	92	23	506			
Na_2HPO_3	anhydrous	Na^+	24	11	222	32.37	1.0273	1.0277
		Na^+	24	11	222			
		HPO_3^{--}	76	8	220			
NaH_2PO_3	1 H_2O	Na^+	24	11	222	35.456	1.0228	1.0218
		$H_2PO_3^-$	80	6	188			
NaH_2PO_4	1 H_2O	Na^+	24	11	222	107.3	1.078	1.073
		$H_2PO_4^-$	96	21	474			

The data given in the preceding table were interpreted as evidence that the presence of hydrogen in associations involving phosphorus and oxygen so increased the forces cementing the units that water was unable to effect dissociation. The interpretation prompted the suggestion that independent of the activities of organisms ionic phosphorus atoms would be limited to alkaline habitats. On the other hand it had been indicated previously that osmosis not only involved dissociation potentialities exceeding those of water but also involved an extrusion of hydrated ions from all membrane-liquid interfaces into the circumambient liquids. It followed that in the event that osmosis could effect the complete dissociation of the solute hydrogen-phosphorus-oxygen radicals ionic hydrated phosphorus atoms not only might be made available internally in organisms possessing vacuolar tissue but also might be made available externally through the activities of such organisms within an aqueous matrix.

Studies of the osmotic behavior patterns of phosphorus—containing solutes in an aqueous solvent involved the accumulation of data from original measurements, since no suitable tabulated data were available. In the absence of phosphite compounds the studies were confined to solutions of common stable chemicals containing sub-ionic P^{+5} atoms. The results obtained have been given in Table 33.

The data given in Table 33 were interpreted as evidence that osmotic potentialities for dissociation exceeded those of water and were adequate to bring about the complete dissociation of three of the four compounds containing associated hydrogen. The surprising item, however, was the indicated failure of osmosis to bring about the complete dissociation of monobasic sodium phosphate, NaH_2PO_4. The observed standard increase in solution volume, 237, was the value subject to calculation for the ions Na^+ plus $H_2PO^-_4$.

Inasmuch as monobasic potassium phosphate, KH_2PO_4, had been evidenced subject to complete osmotic dissociation, and the

Table 33. Data pertaining to the osmotic behavior of some common phosphorus-containing substances in aqueous solution. VT=calculated standard increase in solution volume. O=observed standard increase in solution volume. Other headings as in "Behavior Patterns of Hydration."

Solute	Assumed Ions	W_a	H	W_h	V_h	V_a	D_1	$\frac{D_1}{2}$	$\frac{V_a}{2}$	D_2	U	VT	O
Na$_3$PO$_4$	Na$^+$	24	11	222	200	12	188	94	6	88	3	865	865
	P^{+5}	40	3	94	66	20	46	23	10	13	1		
	O$^-$	12	17	318	306	6	300	150	3	147	4		
Na$_2$HPO$_4$	Na$^+$	24	11	222	220	12	188	94	6	88	2	957.5	957
	H$^+$	4	21	382	365	2	363	181.5	1	180.5	1		
	P^{+5}	40	3	94	66	20	46	23	10	13	1		
	O$^-$	12	17	318	306	6	300	150	3	147	4		
NaH$_2$PO$_4$	Na$^+$	24	11	222	200	12	188	94	6	88	1	1050	237
	H$^+$	4	21	382	365	2	363	181.5	1	180.5	2		
	P^{+5}	40	3	94	66	20	46	23	10	13	1		
	O$^-$	12	17	318	306	6	300	150	3	147	4		
KH$_2$PO$_4$	K$^+$	40	3	94	66	20	46	23	10	13	1	975	975
	H$^+$	4	21	382	365	2	363	181.5	1	180.5	2		
	P^{+5}	40	3	94	66	20	46	23	10	13	1		
	O$^-$	12	17	318	306	6	300	150	3	147	4		
K$_2$HPO$_4$	K$^+$	40	3	94	66	20	246	28	10	13	2	807.5	807
	H$^+$	4	21	382	365	2	363	181.5	1	180.5	1		
	P^{+5}	40	3	94	66	20	46	23	10	13	1		
	O$^-$	12	17	318	306	6	300	150	3	147	4		
K$_3$PO$_4$	K$^+$	40	3	94	66	20	46	23	10	13	3	640	640
	P^{+5}	40	3	94	66	20	46	23	10	13	1		
	O$^-$	12	17	318	306	6	300	150	3	147	4		

Table 34. Data relating to the effect of the potassium ion on the dissociation of the H_2PO_4 radical.

Solutes	Assumed Ions	Wa	H	Wh	Vh	Va	D_1	$\dfrac{D_1}{2}$	$\dfrac{Va}{2}$	D_2	U	VT	O
	K^+	40	3	64	66	20	46	23	10	13	1		
	H^+	4	21	382	365	2	363	181.5	1	180.5	2		
KH₂PO₄	P^{+5}	40	3	94	66	20	46	23	10	13	1		
	O^-	12	17	318	306	6	300	150	3	147	4	2025	2025
+													
	Na^+	24	11	222	200	12	188	94	6	88	1		
	H^+	4	21	382	365	2	363	181.5	1	180.5	2		
	P^{+5}	40	3	94	66	20	46	23	10	13	1		
NaH₂PO₄	O^{--}	12	17	318	306	6	300	150	3	147	4		
	K^+	40	3	94	66	20	46	23	10	13	1		
	H^+	4	21	382	365	2	363	181.5	1	180.5	2		
KH₂PO₄	P^{+5}	40	3	94	66	20	46	23	10	13	1		
	O^-	12	17	318	306	6	300	150	3	147	4	1212	2025
+													
	Na^+	24	11	222	200	12	188	94	6	88	1		
NaH₂PO₄	H_2PO_4	96	21	474	394	48	346	173	24	149	1		

hydrational potentiality of K^+ ion was less than that of the Na^+ ion, there arose the possibility that the hydrated K^+ ion might have potentialities for dissociation. This naturally led to a series of supplementary tests, the results of which have been given in Table 34.

The results given in Table 34 were interpreted as evidence that the hydrated potassium ion, K^+, possessed potentialities for the dissociation of the radical $H_2PO_4^-$, and in conjunction with the data of Table 33 it was indicated that these potentialities were in excess of those exhibited by water and by osmosis. There was the obvious suggestion that a major role of potassium in metabolic processes was that of a catalyst,—a suggestion in keeping with a viewpoint derived from other considerations. Indirectly the data of Table 34 had a bearing on the availability of phosphorus as a metabolite.

DISCUSSION

The results obtained suggested the metabolic phosphorus commonly was present as P^{+5} ions having a weight of 40 in the anhydrous state and hydrated with the 3 H_2O^- units prescribed for ions of that weight by the description of periodic hydrational potentiality. Indirectly there was the suggestion that under conditions tending to be anaerobic P^{+3} ions having a weight of 36 in the anhydrous state and hydrated with the prescribed 5 H_2O^- units might be present.

The results appeared to be of unusual speculative interest. As represented in the data of Tables 29, 30 and 31, which involved simple aqueous solutions, the hydrated P^{+5} ion was present and stable in the absence of hydrogen. The involved aqueous solutions were alkaline. Plants seemingly devoid of vacuoles, such as some bacteria and blue green algae, have appeared to thrive best in alkaline media. On the other hand the development of cellulosic osmotic membranes could be ventured as an important and perhaps essential evolutionary step in the physiological adaptation of organisms to the more acidic waters of land habitats. Certainly such membranes made possible the release and hydration not only of atomic phosphorus ions but also of similarly-bound sub-ionic atoms of other essential nutrient elements, especially when accompanied by potassium as a catalyst.

SUMMARY

Through the use of a description of periodic hydrational potentiality the behavior patterns of some phosphorus—containing solutes were examined. It was found that the stresses involved in an aqueous solvent were sufficient to effect the release of P^{+5} ions from some compounds, and that these ions thereupon became hydrated. It was found further that solute radicals characterized by an association of hydrogen, phosphorus and oxygen remained intact in simple aqueous solution. Some radicals of this nature were found to be subject to complete dissociation by the stresses involved in osmosis, following which the released ionic atoms of the component elements immediately became hydrated. It was found further that the solute potassium ion, K^+, possessed dissociation potentialities in excess of those possessed by water and by osmosis.

Commonly the released phosophorus was the P^{+5} ion of weight 40 in the anhydrous state but immediately acquiring 3 H_2O^- units in hydration as prescribed by the mentioned description. There was the suggestion that under conditions of reduced oxygen supply there might take place a release of P^{+3} ions of weight 36 in the anhydrous state but immediately acquiring the prescribed 5 H_2O^- units in hydration.

CHAPTER 16.   CONCERNING IRON AS A METABOLITE

From many years iron has been recognized as an element essential in the nutrition of plants and animals. Usually small amounts of iron have appeared adequate. In nutrient solutions used for the laboratory culture of organisms the inclusion of a trace of iron commonly has been sufficient to prevent the development of symptoms attributable to iron deficiency.

To the writer the mechanics of trace element activity in metabolism had appeared to be quite outside the realm of approach through studies of osmotic behavior patterns. In the course of some experiments concerned with nitrogen as a metabolite. however, it so happened that a particularly disturbing situation developed. In the book "Behavior Patterns of Hydration" the text of Chapter 8 had been concerned with the osmotic release of nitrogen from ammonium and nitrate radicals when copper sulfate was present in the solution. The involved experimental tabulated data which sustained the text had been obtained with standard iron osmometers, but no special significance had been attached to that fact. However, in conjunction with subsequent further studies relating to the solubility of nitrogen gas in water the earlier experiments were repeated. On this occasion quite by chance some all-plastic osmometers were used. The results obtained were quite different from the results previously obtained with iron osmometers, and indicated that no breakdown of the ammonium and nitrate radicals took place. Data testifying to this development have been given in Table 35.

Taken in conjunction with the results obtained in previous experiments when iron osmometers were used (Tables 18 and 19, Chapter 8, Behavior Patterns of Hydration) there was the suggested possibility that the iron content of the osmometers used in the

Table 35. Data obtained in studies of osmotic behavior patterns carried out with all-plastic osmometers. The observed volumetric increases clearly evidenced no release of atomic nitrogen from the involved radicals. Headings as in previous publication. VT=calculated volumetric increase. O=observed increase.

Solutes	Assumed Ions	W_a	n	H	W_h	V_h	V_a	D_1	$\frac{D_1}{2}$	$\frac{V_a}{2}$	V_1	u	V_2	V_t	O
NH₄Cl	NH₄⁺	24	1	11	222	200	12	188	94	6	88	1	88		
	Cl⁻	32	1	7	158	131	16	115	57.5	6	49.5	1	49.5		
CuSO₄	Cu⁺⁺	62	2	15	332	280	31	249	124.5	15.5	109	1	109		
	S⁺⁶	44	1	1	62	36	22	14	7	11	-4	1	-4		
	O⁻⁻	12	1	17	318	306	6	300	150	3	147	4	558	830.5	830
KNO₃	K⁺	40	1	3	94	66	20	46	23	10	13	1	13		
	NO₃⁻	60	2	16	348	207	30	267	133.5	15	118.5	1	118.5		
CuSO₄	Cu⁺⁺	62	2	15	332	280	31	249	124.5	15.5	109	1	109		
	S⁺⁶	44	1	1	62	36	22	14	7	11	-4	1	-4		
	O⁻⁻	12	1	17	318	306	6	300	150	3	147	4	588	824.5	824

Table 36. Data obtained in studies of osmotic behavior patterns carried out with all-plastic osmometers. Iron compound included with solutes. The observed volumetric increases clearly evidenced a release of atomic nitrogen from the involved radicals.

Solutes	Assumed Ions	Wa	n	H	We	Vh	Va	D_1	$\frac{D_1}{2}$	$\frac{Va}{2}$	V_1	u	V_2	VT	O
KNO₃	K⁺	40	1	3	94	66	20	46	23	10	13	1	13		
	N⁺⁵	24	1	11	222	200	12	188	94	6	88	1	88		
CuSO₄	O⁻⁻	12	1	17	318	306	6	300	150	3	147	7	1029		
FeCl₃	Cu⁺⁺	62	2	15	332	280	31	249	124.5	15.5	109	1	109		
	Fe⁺³	58	2	17	364	314	29	285	147.5	14.5	133	1	133		
	Cl⁻	32	1	7	158	131	16	115	57.5	8	49.5	3	148.5		
	S⁺⁶	44	1	1	62	36	22	14	7	11	-4	1	-4	1513	1513
NH₄Cl	N⁺⁵	24	1	11	222	200	12	188	94	6	88	1	6		
CuSO₄	H⁻	0	1	23	414	414	0	414	207	0	207	4	828		
	Cl	32	1	7	158	131	16	115	57.5	8	49.5	3	148.5		
	Cu⁺⁺	62	2	15	332	280	31	249	124.5	15.5	109	1	109		
	S⁺⁶	44	1	1	62	36	22	14	7	11	-4	1	-4		
FeCl₃	O⁻⁻	12	1	17	318	306	6	300	150	3	147	4	588		
	Fe⁺³	58	2	17	364	314	29	285	147.5	14.5	133	1	133	1808.5	1808

earlier experiments might have played some part in the evidenced release of atomic nitrogen. It was considered imperative, therefore, to carry out a further series of experiments with all-plastic osmometers in which an iron compound was added to the solution. Some results obtained in this series of experiments have been given in Table 36.

A distinctive feature of the experiments whose results have been given in Table 36 was the fact that in the case of the presence of Fe^{+3} ions the activity was not attributable to any stable chemical reaction. This was revealed when in subsequent experiments it was found that the number of Fe^{+3} ions could be reduced to at least one-tenth the number of NO_3^- or NH_4^+ radicals and still bring about the dissociation, though at a slower pace. This feature suggested that although the solute hydrated Fe^{+3} exhibited an unusual affinity for electrons, removable with the O^{--} and H- constituents of the respective NO_3^- and NH^{4+} radicals, a retention of their hold on such constituents was ephemeral, since they repeatedly effected removal. This suggested activity thus became compatible with the behavior patterns designated as catalytic, enzymatic or synergic.

It was obvious that collectively the results rather neatly documented complexity in metabolism. The Cu^{++} ion alone could not bring about the disruption of the nitrogen-containing radicals, since copper nitrate solutions were ineffective inside iron osmometers. This meant that either S^{+6} ions, or O^- ions, or both, as derivatives of the sulfate radical, could break down NO_3^- and NH_4^+ radicals only in the presence of hydrated Fe^{+3} ions. In experiments involving a substitution of hydrated ferrous Fe^{++} ions for hydrated ferric Fe^+ ions no breakdown was evidenced.

It was considered of interest that but for the incidental repetition of experiments originally carried out in iron osmometers in all-plastic osmometers the activity attributable to the ferric Fe^{+3} ion would not have been revealed. The results obtained, however, appeared to be in keeping with the well-recognized influence of small amounts of iron in relation to plant and animal growth.

Notwithstanding the degree of complexity evidenced by the foregoing results there was an enhanced prospect that osmotic behavior patterns might prove useful in analyses of the activity of other trace elements such as zinc, manganese, cobalt and molybdenum. Eventually it should be possible to appraise the various relative intensities of sub-ionic bondages present in molecules. In sub-ionic quadrivalent carbon atoms at least three of the four linkages have appeared subject to easy osmotic disruption. At the present writing all attempts to effect and evidence a disruption of the linkages in CHO ions have been ineffective. There was the prospect that the effect of such substances as toxins, growth substances and antibiotics on the rate of osmotic activity for any solute appropriate as a standard would have significance as an index of a potential relationship to permeability.

CHAPTER 17.   CONCERNING THE PECTATE RADICAL

In botanical science the pectate radical has come to be recognized as a product of plant cellular metabolism primarily and quite generally associated with the elaboration of cell walls.   In chemical science the pectate radical has been characterized as a complex carbohydrate aggregate, bivalent and anionic.   In nature the element calcium has been evidenced as the most common associate of the radical, but other associates have included magnesium, hydrogen and various organic complexes.   In general the radical has appeared to have absorptive and resilient properties in relation to the uptake, retention and release of water, and these properties prompted the presently reported study of its structure, composition and parttern of behavior with respect to hydration.

The validation of a description of hydration as given in the book "Behavior Patterns of Hydration" made available for the first time as authoritative a mathematical basis of approach to problems involving water relationships.   The corollaries of the description differentiated categories of weight and afforded alternative interpretations of important gravimetric data.   Collectively these developments emphasized the value of complete characterizations of the sub-ionic atoms in molecules and of a recognition of change in weight with ionization.   These values were again concretely documented in a supplementary publication dealing with the composition and structure of the chlorophyll molecule.   It has seemed appropriate to direct further attention to this supplementary publication, reprinted herewith as Chapter 8.

Outstanding features of the reported study of the chlorophyll molecule were (1) the projection of a radiately symmetrical chlorophyll *a* molecule consisting of a nucleus surrounded by eight homogeneous units (2) a simple internal modification to yield chlorophyll *b* (3) the projection of a radiately symmetrical carotin molecule containing six of the same homogeneous units projected for chlorophyll *a* and (4) a simple internal modification of carotin to yield xanthophyll. It was obvious that as thus represented the four protoplasmic pigments were inter-related and that carotin could be appraised as a stage in both the synthesis and the degradation of chlorophyll. The four pigments were insoluble in water and dynamic potentialities appeared restricted to resonance. In contrast, the pectate radical appeared to be of promise as a substance having water relationships potentially affording an introduction to the intriguing area of gel formation.

CAROTIN AND THE PECTATE RADICAL

In the mentioned supplementary publication concerning chlorophyll the structural formulas projected for the four protoplasmic pigments involved chemical analyses of composition as reported in conventional chemistry, a complete characterization of all constituent sub-ionic atoms and a recognition of change in weight with ionization. It was natural, therefore, that a study of the pectate radical followed a similar pattern. As the study continued it became increasingly evident that the pectate radical was subject to appraisal as a specifically oxidized form of carotin. Quite unexpectedly a rather simple modification of the previously-projected radiately symmetrical carotin molecule yielded a radiately-symmetrical pectate radical equally subject to correlation with chemical analyses of composition. It was of interest further that whereas the projected transition of carotin to xanthophyll had involved an internal or nuclear oxidative change, the projected transition of carotin to the pectate radical involved an external or peripheral oxidative change. Basically this change was one in which two ionic atoms of oxygen were interposed between central sub-ionic atoms of carbon and hydrogen. The interposition was suggested as having taken place in each of the six circumferential

units of the projected carotin molecule. For a complete characterization of the carotin molecule reference may be made to Figure
5 in the mentioned Chapter concerning chlorophyll. For a
characterization of the oxidative change within each of the six
circumferential units reference may be made to the following
Figure 17.

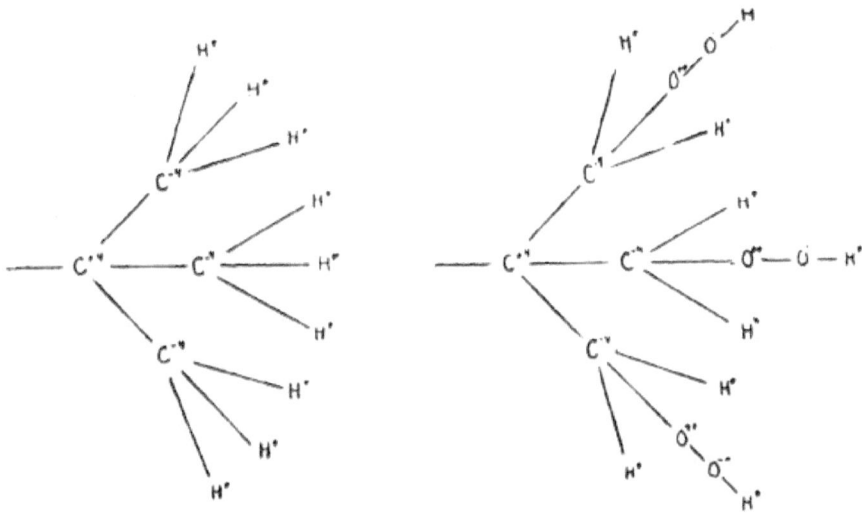

Figure 17. Left: segment of projected carotin molecule, one of six.
Right : segment of projected pectate radical, one of six.

Such a projected modification as that given in Figure 17
naturally arose through correlations. As calculated from the
reference structure of the carotin molecule and the foregoing
Figure 17 the composition of the projected bivalent anionic
radiately symmetrical pectate radical became indicated as $C_{24}H_{54}O_{37}$.
The composition of the pectate radical as derived from conventional chemical analyses of extracted fractions termed pectins
commonly had been represented by the formula $C_{41}H_{60}O_{36}$. The

differences in compostion were obvious, and explanations were in order.

With respect to oxygen it was to be noted that a precise correlation of the projected formula with observational data was possible. For the nucleus of the carotin molecule and the pectate radical a single sub-ionic atom of oxygen was projected as the most likely constituent. From the corollaries of the description of periodic hydrational potentiality such a unit would be 8-valent negative, without weight and hence not subject to gravimetric appraisal. Of the 36 sub-ionic atoms of oxygen projected as contained in the periphery of the pectate radical, 18 were 2-valent positive and 18 were 2-valent negative. Although these units would have different prescribed combining weights, collectively they would have an average weight of 16 and be subject to gravimetric appraisal on that basis. Insofar as oxygen was concerned, therefore, the conventional chemical analyses were subject to a satisfactory correlation with the projected radiately symmetrical structure for the pectate radical.

With respect to hydrogen it was to be noted that only an approximate correlation of the projected composition with observational data was indicated, the conventional chemical analyses having been interpreted as denoting 60 hydrogen atoms in contrast to the prescribed 54 hydrogen atoms. However, it was to be noted that in the event that pectic acid instead of the pectate radical was present in chemical analyses there would have been involved two supplementary sub-ionic atoms of hydrogen with univalent positive ionization, each having a combining weight of 3. Under such circumstances gravimetric data subject to interpretation as evidencing a composition apparently involving 60 hydrogen atoms would not be unusual.

With respect to carbon it was to be noted that the conventional chemical gravimetric data had been interpreted as evidencing 41 atoms, whereas the projected radiately symmetrical pectate radical contained only 24 atoms. Yet in the case of the pectate radical, as in the case of chlorophyll and of carotin, conventional analytical procedures in chemical science did not involve any complete characterization of the sub-ionic atoms nor any recognition of change in weight with ionization. Carbon weighed as a constitutent of carbon dioxide was to be appraised as uniformly 4-valent positive, with a combining weight of 16. In the projected radiately-symmetrical pectate radical, however, only 6 of the sub-ionic carbon atoms were 4-valent positive with a combining weight of 16, and 18 were 4-valent negative, with a combining weight of 8. Under such circumstances 18 carbon atoms weighed half as much in the pectate radical as in the dioxide. In absence of an allowance for change in weight with ionization the number of carbon atoms would be subject to calculation as $6 + (2 \times 18)$, or $6 + 36$, or 42, a value approximating the 41 carbon atoms interpreted as present from chemical analyses. Changes incident to the degradation of the pectate radical or of pectic acid readily might account for the derivation of the value 41 instead of 42.

It became obvious that to a remarkable degree the conventional chemical anlyses of the pectate radical or of pectic acid sustained the projected structural and compositional formula. In so doing they again emphasized the importance of two circumstances which have appeared to have handicapped chemical science: (1) the failure to fully characterize sub-ionic atoms and (2) the failure to recognize change in weight with ionization.

The chemical reaction potentialities of the projected pectate radical were localized in the nucleus and presented no unusual characteristics. The bondages of the six homogeneous circumferential units on the 8-valent nuclear sub-ionic oxygen atom left the two unsatisfied negative charges of a conventional bivalent anion.

The interpretation of the pectate radical as a rather simple oxidative modification of carotin appeared to be of special interest because of the difference between the two substances in their relation to water. On the one hand, carotin was insoluble in water. On the other hand the behavior pattern of the pectate radical with respect to water was quite in contrast and was subject to appraisal as supplying a potential introduction to the intriguing area of gelation. On this account special attention was given to the behavior of the pectate radical in water.

At the outset it was to be noted that in all probability any gelation phenomenon which involved a substance widely distributed in the plant world would be mediated by hydrational bondages. Yet it was clear from the description of periodic hydrational potentiality that the intact pectate radical was much too heavy an ion to hydrate. It had been noted further, as particularly indicated in a publication dealing with color and hydration, that in simple solution a number of oxygen-containing radicals had been evidenced as having undergone partial or complete dissociation. Collectively these considerations made it seem altogether likely that gelation involved a partial or complete disruption of internal chemical bondages followed by an incomplete satisfaction of the potential hydrational bondages of the resultant ions.

It was a relatively simple matter to assume a type of denaturation for the projected pectate radical and then to calculate for the released solute ions the weight ratios which would characterize the earliest attainment of a condition in which no free aqueous solvent was present. For a partial dissociation effecting the separation of the six circumferential units from the nucleus, each of the six units would have a calculated weight of 158 and a prescribed potentiality for taking on 13 H_2O^- units in hydration. The single nuclear ion would have a prescribed potentiality for taking on 23 H_2O^- units in hydration. Under these conditions on a weight basis in a mixture of the pectate radical and water, disregarding associates, it would require 35.2% pectate radical to effect an absence of free solvent, It was obvious that such proportions were not involved in the behavior pattern of the pectate radical in water, since the formation of a gel required far less material. On the assumption that gelation involved a complete dissociation of the pectate radical and the hydrational potentialities of the resultant ions it became of interest to calculate the weight ratios which would characterize the earliest attainment of a condition in which no free solvent was present. The calculations for the projected pectate radical alone, disregarding associates, indicated that it would require 2.48%, — a value seemingly in line with observational data. In relation to practical and industrial aspects it has come to be recognized that gelation phenomena involving "pectins" are influenced by temperature, by the source of the material and by the techniques of the extraction and purification processes. It was of interest further that an assumed complete dissociation of the projected pectate radical, followed by a complete hydration of the resultant ions would effect a calculated 7600% increase in volume of the solute. It was to be recognized also that substantial hydrational bondages might well establish gelation previous to the complete exhaustion of the supply of solvent and represent interstitial occlusion.

Collectively the foregoing considerations appeared to justify the serious appraisal of the pectate radical as a radiately symmetrical oxidative derivation of carotin having the composition $C_{24}H_{54}O_{37}$. On this account a structural formula for the radical has been included as Figure 18.

DISCUSSION

The foregoing appraisal of the pectate radical as a specific oxidative derivative of carotin seemed to have several aspects of interest. The correlation of its projected compositional formula with conventional chemical analyses gave emphasis to the importance of a recognition of change in weight with ionization. The

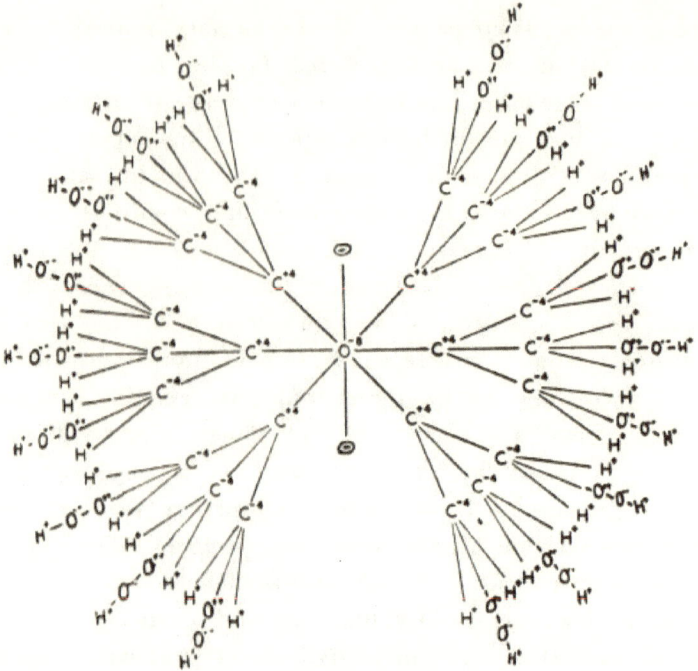

Figure 18. Projected structural formula for the pectate radical.

projected structural formula gave emphasis to the importance of a complete characterization of sub-ionic atoms.

The basic features shared by carotin and the pectate radical were such as to suggest metabolic interchanges. The contrasting insolubility of carotin in water and the relation of the pectate radical to gelation suggested that the interposed oxygen of the latter contributed in an important way, along with heat, to its complete dissociation with the attendant involvement of hydrational bondages. There was thus the added suggestion that the oxygen chemical bonds were weaker and that the released oxygen itself took part in furthering the breakdown of the aggregation. These suggestions in a speculative manner were of potential significance in dynamic aspects relating to the metabolism of alveolar breathing and intracellular respiration.

Within the area of plant physiology the pectate radical as an effluent or extrusion appeared to possess, by virtue of its susceptibility to dissociation, hydrational absorptive potentialities for water both independent of and related to the cellular metabolism, serving as an intercellular resilient cushioning, a reducer of cell-on-cell growth frictions and as a potential source of nutrient ions.

SUMMARY

The pectate radical was projected and evidenced as an oxidative derivative of carotin. Its pattern of behavior with respect to gelation was interpreted as attributable to the hydrational bondages of its sub-ionic atoms following their release from chemical bondages within an aqueous medium or matrix.

CHAPTER 18· CONCERNING THE ASCENT OF SAP

For many years a satisfactory explanation of the ability of tall trees to conduct sap to their uppermost branches has been sought without success. With the departure of vitalistic forces from plant physiology it became obvious that explanations of all physiological "phenomena" were to be sought in the areas of physics and chemistry. One thing tnat was not obvious, however, was the fact that physicists and chemists had not provided adequate basic information. Of special importance was knowledge concerning solutions. The relation of solutes to an aqueous solvent was unknown. As indicated previously, there were only theories of solution,

In the natural confusion associated with this lack of information students often read that water was being sent through osmotic membranes into a more concentrated solution by diffusion because there was less water in the more concentrated solutions. They often read also that the columns of sap within trees were under tension and were kept intact by cohesive forces. Water therefore was reported as acting like a gas with free independent units in osmosis, but as acting like a liquid with bonded units inside trees. Commonly there was no particular concern that the inconsistency documented ignorance.

In 1964 the publication of the book "Behavior Patterns of Hydration" supplied a background for new approaches to old problems. This background had the distinct advantage of being mathematical. It further involved changes in appraisals of osmosis, the density of solutes and interdiffusion. Subsequent studies based on this background included such subjects as the solubility of gases in water, chlorophyll, color and hydration, and turgor.

The text of the present chapter represents a new approach to the old problem of explaining the ascent of sap in tall trees and is based upon principles contained in the mentioned publication. A major disturbing aspect of the problem of the ascent of sap has been a recognition of the fact that with the creation of a vacuum under conditions of external atmospheric pressure a column of water of a height no greater than about 28 feet has been attained. Investigators have not lacked resourcefulness in accounting for the ascent of sap to heights of several hundred feet within trees, but no explanation has been noteworthy as widely satisfactory. Even as a cooperative venture with a push attributed to root pressure, a pull attributed to evaporation and the intermediate services of cohesion and adhesion the ascent has not been an easy one to verify en route to plausibility.

Within appropriate limits the gas laws have been characterized as the most perfect descriptions relating to natural behavior patterns. Commonly the word "perfect" has not been used elsewhere in science. These gas laws had a direct influence upon the appraisal of solutions in that since one gram molecule of any gas under standard conditions of temperature ($0°C$) and pressure (760 mm) occupied a volume of 22.4 liters, one gram molecule of any solute in one liter of aqueous solution was considered as having a concentration 22.4 times as great. This concentration, and others related to it in the event of dissociation, became designated with the term "atmospheres"—a natural but somewhat confusing terminology. When the pressures which arose incident to osmotic activity were measured it was found that at least in some instances these pressures were subject to direct correlation with atmospheric pressure and with solute concentrations appraised in terms of atmospheres. Thus the term "atmospheric pressure" came into use to denote the actual or calculated pressure which a solute concentration might evidence or be expected to evidence in osmosis. Again it was a natural but somewhat confusing terminology.

With respect to osmotic activity evaluated in terms of pressure it appeared that in general all solute units, regardless of the diverse natures, had the same effect, in which case the solutes truly behaved as gases, not only as independent units but as uniformly-acting units. However, in the writer's researches osmotic activity was evaluated in terms of mediated increases in solution volume under standard conditions of temperature (25°C) and pressure (760 mm). These researches yielded results of an entirely different order, and evidenced different specific hydration values for solutes, the values being conditioned entirely by the weight of the involved ion or ions in the anhydrous state. Thus through a standardization of osmotic activity in a manner analogous to the standardization of gas activity, and involving ultimate solution volumes rather than pressures, it was found that osmosis was a most valuable accessory tool in documenting the validity of the description of periodic hydrational potentiality. Such a documentation was included in the mentioned book.

A second discovery of special importance in relation to the problem of the ascent of sap was that osmotic membranes possessed more dissociative ability than the aqueous solvent and brought about a complete or partial breakdown of numerous solute radicals. Allied with this development and essential in the interpretation of osmotically-mediated increases in solution volume was the discovery of the relation of hydration to the density of solute units.

A third discovery of special significance in relation to the ascent of sap was that non-electrolytes became ionic in aqueous solution, their failure to conduct an electrical current being attributable to the presence of both positive and negative charges. Their hydrational potentialities were evidenced as conditioned by their weight in the anhydrous state, as was the case with electrolytes.

In this chapter several assumptions will be made with regard to the general conditions basically involved in the ascent of sap. One assumption is that glucose is either a major carbohydrate involved or will serve as a satisfactory representative of the carbohydrate content of cell sap. Another assumption is that within plant tissues glucose is cyclic and of the composition $C_6H_6O_6$. The usual compositional formula for glucose has been $C_6H_{12}O_6$, but from the corollaries of the description of periodic hydrational potentiality the weight of the neutral hydrogen atom is 2 instead of the 1.0 or 1.008 conventionally assigned to the element, and there is ample evidence to sustain the validity of the corollaries. Another assumption is that glucose in aqueous solution becomes ionic with three positive and three negative charges. This assumption is in keeping with the behavior patterns of other carbohydrates studied and with the fact that solubility in water invariably has involved the presence of ions.

On the basis of the foregoing assumptions the weight of the glucose ion $C_6H_6O_6^{+3-3}$ was calculated as 180. From the description of periodic hydrational potentiality an ion of weight 180 would have a prescribed ability to take on and hold $2H_2O^-$ units of weight 18. The weight of the ion so hydrated would then be $180 + 36$, or 216. As reported in the mentioned book the indicated relationship of hydration to solute density was $d = 1 + \dfrac{wa}{wh}$, when d = density, wa = weight in the anhydrous state and wh = weight in the hydrated state. The density of the solute hydrated glucose would be calculable as $1 + \dfrac{180}{216}$, or 1.833, and 833 grams would dissolve in water to yield a liter of a saturated glucose solution. This value approximates the order of solubility given in tables of physical and chemical data. It seemed appropriate to conclude that in aqueous solution the glucose was present in the assumed ionic state and hydrated to the indicated prescribed extent.

The osmotic behavior of an aqueous solution of glucose as thus represented was a simple matter to calculate. From the reference data of Table 80, on page 50 of the mentioned book the standard behavior pattern for a liter of a gram molecular solution of glucose would be to undergo an osmotically-mediated decrease of 31 milliliters in solution volume. While it has been well known that a glucose solution has a potentiality for effecting osmotic pressure and an osmotically-mediated increase in solution volume, the significance of this increase in solution volume was not known. It could not become known in the absence of a description of hydration.

Following the assumed osmotically-mediated disruption of the carbon-to-carbon linkages the glucose ionic molecule became subject to representation as having given rise to six CHO ions, three positive and three negative. These ions might or might not involve a double bonding of the sub-ionic oxygen atoms. To satisfy curiosity it would be pleasant to know their precise status, but the matter was of no importance from the standpoint of ionization and hydration: in either case the average anhydrous weight for the six CHO ions would be 30. From the reference data of Table 8A, on page 49 of the mentioned book, the standard osmotic behavior for a liter of a gram molecular solution of an ion of anhydrous weight 30 would be to effect an increase of 59 milliliters in solution volume. For a six gram molecular solution the prescribed increase would be 354 milliliters. For strict mathematical accuracy the total increase calculated for three positive ions and three negative ions is not quite the same as the increase calculated for six ions of average weight. However, the observed standard increase in solution volume for a molar solution of glucose was approximately 354 milliliters. It seemed not only appropriate but necessary to conclude that osmosis had broken the carbon-to-carbon linkages of glucose molecules originally intact and cyclic. Such a conclusion was entirely in keeping with results obtained in studies of the behavior patterns of other solutes involving carbon-to-carbon linkage.

The developments up to this point may seem to have had only an obscure subtle potential bearing upon the problem of the ascent of sap. They supplied a basis, however, for some considerations and calculations of interest in relation to the problem. One of the considerations was the fact that the CHO ions evidenced as resulting from the osmotic disruption of the carbon-to-carbon linkages had an average ability to take on and hold 8 H_2O- units in hydration, and that this ability was an approximation of the ability of projected atmospheric ions of nitrogen and oxygen to take on H_2O- units in hydration. The abilities would precisely counterbalance if there were equal numbers of N_2^+, N_2^-, O_2^+ and O_2^- ions present in the atomosphere, since these ions have an average weight of 30. Non-ionic molecules by prescription would have no hydrational potentialities, Physiologically it followed that the presence of glucose in leaves under conditions of osmotic activity would confer a substantial degree of resistance to desiccation attributable to the removal of H_2O- units to effect the hydration of atmospheric ions.

Yet by calculation a solution of CHO ions would become saturated when a ratio of 17.4 grams to a liter of aqueous solution had been attained. Here it could be ventured that translocation appropriately might involve CHO ions and that any subsequent dehydration might result in the formation of cyclic glucose.

It became of interest to calculate on each of two bases, pressure and solution volume, the theoretical heights to which a 1% solution of cyclic glucose could elevate a column of liquid against gravity. On the basis of pressure a saturated glucose solution would contain 45.46% glucose, or 8.48 gram molecules. A 1% solution of glucose would contain .186 gram molecules, or have a concentration of 4.16 atmospheres. If the solute units remained intact such a solution theoretically might be expected to exert in osmosis a pressure sufficient to elevate a column of water to a height of 58.24 feet. From the indicated behavior patterns of osmosis, however, there would be a decrease rather than an increase in solution volume, as previously noted. Yet in actual osmotic activity the carbon-to-carbon linkages had been evidenced as having become broken, and the resultant ions as having become hydrated immediately to the full extent of their prescribed hydrational potentialities. On the basis of pressure a 1% solution of cyclic glucose following osmotic dissociation would contain 1.116 gram molecules, or have a concentration of 25 atmospheres. From the description of osmosis given in the mentioned book it was clear that only one-half the number of solutes taking part in osmosis were involved in the osmotically-mediated increases in solution volume. It would be natural and perhaps necessary to conclude that this number would be involved in the exercise of pressure. Under the projected conditions the 1% solution of glucose would have in osmosis a pressure sufficient to elevate a column of water against gravity to a height of 350 feet. On the basis of an osmotically-mediated increase in solution volume following a disruption of carbon-to-carbon linkages there would be a 56% increase in the volume of the solution as compared with the aqueous solvent. Under these circumstances it became obvious that theoretically the translocation of enough photosynthate from the leaves to the roots to produce a 1% solution of cyclic glucose could account for the elevation of sap to the tops of the tallest trees. '

CHAPTER 19. HYDRATION AND CHEMICAL SCIENCE

In recent decades chemical science has become well aware of hydration as an important aspect of the behavior patterns of solutes and of solids crystallizing from aqueous solutions. In the absence of a satisfying description of hydration, however, the appraisals of hydration by chemists have appeared abstruse and unproductive beyond the frontiers of philosophical resourcefulness. The situation has served to prolong the life of irregular fractional atomic weight values which, although often as greatly treasured as imagined supernatural gods, have seemed to be equally doomed to eventual extinction. Unfortunately the utmost in meticulous measurement failed to insure an appropriately precise interpretation of the resulting gravimetric data of chemical science. The description of periodic hydrational potentiality and its involved corollaries defining categories of weight are basic contributions to chemical science. These contributions at present may well seem unacceptable. They represent departures from conventional interpretations and document errors in the precepts of a discipline which has accomplished much and has attained an imposing aura of prestige. Yet biologists are free to make their own evaluations within the areas of physics and chemistry, for it is within these areas that they must find explanations of the processes involved in the behavior patterns of plants and animals. The description of periodic hydrational potentiality and its involved corollaries defining categories of weight are far more important contributions to biology as an emergent young science than to chemistry, for hydrational potentiality has been an intimate and indispensable adjunct to plant and animal evolution throughout the development of the organic world.

In the recent book entitled "Behavior Patterns of Hydration" the specific gravity of aqueous solutions was used quite extensively in documenting the validity of a satisfying description of periodic hydrational potentiality. The documentation involved excellent orders of agreement between calculated values for specific gravity, based on the description, and observed valves obtained during the later part of the previous century by investigators world-renowned for devotion to detail. It was recognized and emphasized that no more trustworthy criterion of authenticity could be projected than the involved order of agreement. For good measure the description of hydration was validated further by osmotic data.

As an aftermath of validation it became evident that the specific gravity of aqueous solutions was an attribute having accessory values, and one of these values consisted of significance with respect to the characterization of the components of chemical compounds. Historically, in the writer's varied researches relating to hydration the initial supplementary significance of observed values for the specific gravity of aqueous solutions became apparent in conjunction with a study of sulfates. In aqueous solutions of numerous sulfates a satisfactory order of agreement between calculated and observed values for specific gravity was obtained when the sulfate radical was projected as having remained intact. This was entirely in accord with the viewpoints of conventional chemistry. However, in aqueous solutions of copper sulfate a satisfactory order of agreement between calculated and observed specific gravity values was obtained only upon the assumption that a complete dissociation of any projected sulfate radical had taken place. Later on in conjunction with the study of color and hydration reported in Chapter 9 aqueous solutions of copper sulfate were evidenced by another index as consisting of atomic ions exclusively. It appeared that the behavior patterns of projected solute sulfate radicals were conditioned by aasociation : with copper there was dissociation, whereas with most elements they remained intact, This diversity of behavior became subject to further representation in relation to chemical composition by an examination of data for the specific gravity of aqueous solutions of uranyl nitrate, for in such solutions the uranyl radical was evidenced as having become completely dissociated, whereas the nitrate radical was evidenced as having remained intact, In the data of the following Table 37 the observed values are those of the Int. Crit. Tables, Vol. 3.

Table 37. Data relating to the specific gravity of uranyl nitrate in aqueous solution.

Electrolyte	Assumed Ions	Wa	H	Wh	$\dfrac{Wa}{Wh}$	Average $\dfrac{Wa}{Wh}$
	U^{+6}	196	0	196	1.000	
	O^{--}	12	17	318	0.0377	.284
$UO_2(NO_3)_2$	O^-	12	17	318	0.0377	
	NO_3	60	16	348	0.1724	
	NO_3	60	16	348	0.1724	

%	Gms/h	÷284	÷1.284	Ml Solvent	Calc. Sp. Gr.	Obs. Sp. Gr.
10	109.1	383	298	702	1.085	1.091
16	184	648	505	495	1.143	1.149
20	239	844	657	343	1.187	1.194
26	330	1163	906	94	1.257	1.269
30	397.5	1400	1090	90	1.310	1.322
36	506	1783	1389	389	1.394	1.406
40	586	2063	1608	608	1.455	1.466
46	720	2530	1970	970	1.560	1.567
50	825	2910	2265	1265	1.645	1.649

The evidenced complete dissociation of the uranyl radical brought forward for further consideration the attributes of the uranium atomic ion U^{+6}. The anhydrous weight of 196 was outside the weight range of periodic hydrational potentiality, whose upper limit was 184, the weight of the neutral uranium atomic atom. The data evidenced that the U^{+6} ion remained anhydrous, as prescribed. The integrity of all of the involved assumptions and values was evidenced. This again confirmed that the specific gravity of aqueous solutions in conjunction with the indicated description of hydrational potentiality was an attribute of service following chemical analysis as an index not only of the correctness of the derived composition but also of the nature and arrangement of the involved sub-ionic constituents. Inasmuch as some of the arrangements were subject to modification in osmosis the attribute was appraised as of special significance in biology, but the analysis of such modifications likewise appeared to be a matter of considerable promise in chemistry. It was also made evident that when anhydrous ions were included in the aqueous solution the specific gravity of the solution was conditioned by the averages of the ratios of the ions present, not by the ratios of the summation values. This interesting feature previously had been documented in the data of Tables 34 and 36 of the preceding volume. One might venture or conclude, therefore, that in mixtures of hydrated and anhydrous solute ions there was a modified adjustment somewhat modifying the individuality and independence represented in the Kohlrausch Law of the Independent Migration of Ions. Perhaps the most important feature of the data of Table 37, however, was the simple fact that the observed specific gravity values attested the integrity of such ionic weight values as $0^{--}=12$ and $U^{+6}=196$, for any departure from the assumed weight values would have distorted the involved mathematical integration of all the involved data.

Another accessory value for the specific gravity of aqueous solutions in relation to chemical science was that concerned with the internal arrangements in dimorphic salts. In the International Critical Tables, Vol. 3, there was listed a series of specific gravity values for aqueous solutions of "violet" and "green" chromium sulfate, $Cr_2(SO_4)_3$. For each chemical the violet and green types had the same chemical composition, yet there were differences in specific gravity as well as in color. The objective of the studies herewith reported was to determine the specifie significance of the indicated colors. With respect to the violet type of chromium chloride the developments were as indicated in Table 38.

Table 38. Data pertaining to the specific gravity of aqueous solutions of the violet type of chromium chloride, $CrCl_3$.

Electrolyte	Assumed Ions	Wa	H	Wh	$\dfrac{Wa}{Wh}$
$CrCl_3$	Cr^{+3}	54	19	396	150 = .1724
	Cl^-	32	7	158	— —
	Cl^-	32	7	158	870
	Cl^-	32	7	158	
		150		870	

%	Gms/Liter	÷1724	÷1.1724	Me Solvent	Calc. Sp. Gr.	Obs. Sp. Gr.
1	10.076	58.40	49.8	950.2	1.0086	1.0076
2	20.332	117.80	100.48	899.52	1.0173	1.0166
4	41.396	240	205	795	1.035	1.0349
6	63.210	366	312	688	1.054	1.0551
8	85.792	497	424	576	1.073	1.0751
10	109.170	634	540	460	1.094	1.0917
12	133.680	774	660	340	1.114	1.1114
14	158.424	920	784	216	1.136	1.1316

The data given in Table 38 were interpreted as evidence that in aqueous solution the violet type of chromium chloride contained atomic ions exclusively. In contrast to this situation it was found that the green type of chromium chloride in aqueous solution was subject to representation as given in Table 39.

Table 39. Data pertaining to the specific gravity of aqueous solutions
of the green type of chromium chloride, $CrCl_3$.

Electrolyte	Assumed Ions	Wa	H	Wh	$\frac{Wa}{Wh}$
$CrCl_3$	$CrCl.^{+2}$	86	3	140	$\underline{150} = .3283$
	Cl^-	32	7	158	456
	Cl^-	$\underline{32}$	7	$\underline{158}$	
		150		456	

%	Gms/Liter	÷.3283	÷1.3283	M Solvent	Calc. Sp. Gr.	Obs. Sp. Gr.
1	10.071	30.63	23.08	976.92	1.0075	1.0071
2	10.314	61.9	46.7	953.3	1.0152	1.0157
4	41.328	125.8	94.6	905.4	1.0312	1.0332
6	63.06	192	144.8	855.2	1.0472	1.0510
8	85.528	260.3	196	804	1.0643	1.0691
10	108.76	332	250	750	1.082	1.0876
12	132.78	404	304.2	659.8	1.0998	1.1065

The data given in Table 39 were interpreted as evidence
that in aqueous solution the green type of chromium chloride
contained hydrated ions which included radicals. As thus
interpreted the results suggested that a similar set of differences
would be found to characterize solutions of the violet and green
types of chromium sulfate. The results obtained have been given
in Tables 40 and 41.

Table 40. Data pertaining to the specific gravity of aqueous solutions of the violet type of chromium sulfate $Cr_2(SO_4)_3$.

Electrolyte		Assumed Ions	Wa	H	Wh	$\dfrac{Wa}{Wh}$
$Cr_2(SO_4)_3$		Cr^{+3}	54	19	396	$384 =$ .0802
		Cr^{+3}	54	19	396	$\overline{4794}$
	3	S^{+6}	44	1	62	
	12	$O-$	12	17	318	

%	Gms/L.	÷0802	÷1.0802	M Solvent	Calc. Sp. Gr.	Obs. Sp. Gr.
2	20.382	254	235	765	1.019	1.0191
4	41.580	518	480	520	1.038	1.0395
6	63.624	794	734	166	1.06	1.0604
8	86.536	1080	1000	0	1.08	1.0817
10	110.34	1375	1272	-272	1.103	1.103
14	160.804	2008	1857	-857	1.151	1.1486
18	215.2	2680	2480	-1480	1.20	1.1966
22	274	3420	3162	-2162	1.258	1.2479
26	338	4230	3915	-2915	1.315	1.3032

Table 41. Data pertaining to the specific gravity of aqueous solutions of the green type of chromium sulfate, $Cr_2(SO_4)_3$.

Electrolyte		Assumed Ions	Wa	H	Wh	$\dfrac{Wa}{Wh}$
$Cr_2(SO_4)_3$		$CrSO_4^+$	146	19	488	
		$CrSO_4^+$	146	19	488	$384 =$ .166
						$\overline{2310}$
		S^{+6}	44	1	62	
	4	$O-$	12	17	318	

%	Gm|L.	÷166	÷1.666	M Solvent	Calc. Sp. Gr.	Obs. Sp. Gr.
2	20.344	122.3	105	895	1.0173	1.0172
4	41.431	250	214.5	785.5	1.0355	1.0358
6	63.306	381	326.5	673.5	1.0545	1.0551
8	86.008	518	444	556	1.074	1.0751
12	133.9	806	692	308	1.114	1.1172
16	186	1120	961	39	1.159	1.1618
20	241.82	1455	1250	-250	1.205	1.209
24	302	1820	1560	-560	1.260	1.2594
28	368	2220	1903	-903	1.317	1.3125

The data given in the preceding four Tables were considered to be reliably significant by virtue of correlations and consistency. As noted, a complete dissociation of the sulfate radical incident to aqueous solution had been established for copper sulfate, on which account a similar development with respect to chromium sulfate was not surprising. The unusual feature of the data was the indicated presence of such radicals as $CrCl^{++}$ and $CrSO_4^+$ in aqueous solution and their obvious correlation with the green color of the crystalline salts. This correlation made it seem altogether likely that the radicals had pre-existed as sub-ionic entities in the salts. Conversely, the obvious correlation of a violet color with absence of solute radicals made it seem altogether likely that the sub-ionic entities in the salts had been exclusively atomic.

Collectively the data of the four Tables appeared suggestive beyond the immediate indicated correlations. The involvement of the element chromium introduced the attribute of colors, but it seemed plausible that in the absence of color difference analogous variant forms for other salts commonly would not be recognizable. This possibility appeared to enhance the value of the specific gravity of aqueous solutions as a check not only on gross chemical composition but also on internal arrangements of sub-ionic constituents.

CHAPTER 20.   CONCERNING GOALS

During recent decades the experiment-allied curiosity which brought in the age of science has extended into areas involving history, involving behavior patterns of humans in ages long before written languages, in ages representing for all self-styled civilized people a common emergence from the depths of ignorance and superstition. It was the same ignorance and superstition which today characterize the lives of primitive tribes replete with witch-doctors securing domination and exploitation through purported collusion with imaginary supernatural forces, spirits, gods. This ignorance had done much to account for the rise of human shepherds and the allied subservience of human sheep. An inevitable goal of biological science is to orient humans within a natural world of surpassing beauty and challenge. Fortunately this natural world contains many behavior patterns for human living which far transcend those transcribed from the experiences of ancient nomadic tribes beseeching imagined supernatural gods in the tradition of an ancestral caravan of predecessors extending back into the beginnings of human wonderment, awe, ignorance and fear. Extending into the beginnings of the spirit world of ancient and modern aborigines.

It has been said that there are none so blind as those who do not wish to see, and none so ignorant as those who do not wish to learn. The responsibilities of biological science are clear but awesome. Because of the extent of domination and exploitation exercised by contemporary organized religions the intellectual evangelism which is education will have unwilling students. It will need courage as well as dedication, for as present only Russia has had leaders intelligent enough and courageous enough to protest continued addiction to heirloom obsessions perpetuating traditional ignorance. Other nations point with scorn and cling to ancient gods.

The truth seems to be that when plants and animals were domesticated, that is, dominated and exploited, the animals included the meek and less nimble-witted of the humans, the unthinking humans content to be followers, the human sheep of the human shepherds. Biological science is more than a marriage of botany and zoology, neither of which has ever looked much beyond form and physiology in their respective disciplines. Biological science envisions the natural organic world as a dynamic spectacular, unified by the fire that is life and bright with the prospects for the further evolutionary enrichment of life. These prospects go beyond the procurement of a more intimate knowledge of nature. They include the challenge and the responsibility of finding in nature patterns of behavior which for humans will gloriously supplant the long treasured but fictitious viewpoints regarding the supernatural. Nature abounds in such patterns. The amazing thing is that the patterns have remained for so long a time unrecognized. Seven thousand years ago Egyptians spent much time, effort and expense planning for life after death,— and today many humans still believe that utter nonsense, thoughtlessly spawned in ignorance and wistful conceit. There are people living today who still worship the mythical gods and goddesses of Mt. Olympus. One might well rejoice and take pride in being human, but one might well be cautious in deprecating the creatures designated as "dumb animals." Humans not only have no monopoly of brains: they do not always use what they have. In some ways humans have been better at aping than the apes themselves: aping viewpoints which preceded any historical genesis, and continuing to cherish them as birthrights. Unfortunately some of these viewpoints have sought to impair the natural delight of having been born into a race of humans subject to characterization as a special kind of ape.

Genealogy is a science which is of great potential service in social illumination. Although it is limited by historical records insofar as individual human personalities are concerned there are other ways to derive prehistory information. The use of words subject to translation as "shepherd" and "sheep" in early historical records clearly has denoted an antecedent pastoral background. And other words idealizing green pastures and still waters have indicatad a common incidence of unfavorable conditions of pasturage. Yet nomadic humans gaining a livelihood through the exploitation of domesticated sheep represent innumerable centuries of change from the status of aboriginal humans such as those persistent even at the present time in regions more or less isolated from the main streets of civilization. The contemporary aboriginal humans therefore afford genealogical clues to developments which precede the attainment of the pastoral state. The living witchdoctors of primitive tribes unintentionally but none the less definitely evidence the true origin of notions regarding supernatural forces. As indicated previously, it will be a difficult but inevitable goal of biological science to supplant these notions with a far more satisfying knowledge of nature. Humans being the way they are, with none so ignorant as those who not wish to learn, many shepherds will join their sheep in bleating.

Currently one of the goals of biological science is to institute life within the laboratory. Such an accomplishment would supply further evidence that life on this earth originated as a fortuitous aggregation of solutes in an aqueous solvent. Already there is abundant evidence of such an origin. The "new" life would be an interesting subject for the study of environmental factors in relation to development, and subsequently for the study of heredity. Of seemingly greater importance at the present time, however, is the matter of a broader dissemination of the knowledge already acquired. It is a sad commentary on our times that uneducated people continue fanatic evangelism for this or that fiction, or against magnificent natural evolution, and thus promulgate ignorance and delusions. It is a sad commentary that today's world contains so many humans who are content to be the sheep of shepherds who often are almost completely unaware of the significance of scientific knowledge attained in recent years. These are sad commentaries because the natural world, as noted previously, has far better patterns for human living than any prescribed in the writings of ill-informed ancients.

From biological science students may obtain a zestful appreciation of the joy associated with the opportunity to carve out a career in the area of personal potentialities, to enjoy life while doing so, and to enjoy also the satisfaction of making some contribution, however small, toward a better world for their contemporary associates and those who come afterward. In the final analysis much of this is what a leaf does, what every cell of a leaf does, what every cell of a plant or animal that ever lived has done. In plant and animal tissues the cells are organized, but appear to act as though in a sort of league not only for mutual individual benefit but also for the benefit of the organism of which the tissue is a part. Life for some cells may be a tranquil thing, for others a feverish thing, but there is adaptation and coordination and cooperation. In these ways nature is wonderful. What is not as wonderful is the human weaving of the tapestry called civilization. One may attribute flaws to innocence, to ignorance, to greed, to a desire for domination and exploitation. Whatever the explanation, when there is no room for joy in life there is a flaw somewhere. All young animals are playful, and it is a sign of progresss that fewer adult humans now hold that everything or anything affording pleasure is sinful. If one holds that there is no joy in plants it is a certainty that he has never studied biology to the extent of watching sperm cells trying to fertilize egg cells in brown algae. For the beginning student in biology there is the fun of watching an amoeba have fun, and the fun of realizing that both the observer and the observed are a part of a magnificent natural evolution. As a matter of fact the observer is a very highly privileged part of natural evolution, and biological science stands ready to help one make the very most of the part.

Appendix A – Matilda Moldenhauer Brooks, *Advancing Frontiers of Plant Science* 16 (1960), 51-70: "Atomic Energy Levels Explaining Photosynthesis"

ATOMIC ENERGY LEVELS EXPLAINING PHOTOSYNTHESIS

Matilda M. Brooks

Department of Physiology University of California, Berkeley, California.

SUMMARY

The force controlling photosynthesis is the cycle of day and night acting upon electrons which absorb energy in the light and lose it in darkness. The key is the magnesium atom of chlorophyll, and its specific wave lengths which absorb ultra violet light. The energy so absorbed results in the attraction of electrons from the oxygen atoms of water and carbon dioxide. The loss of the first valence electron from the oxygen atom of water and from the first oxygen atom of carbon dioxide, produces an unpaired seventh electron. Such two atoms form a covalent bond and unite to produce the molecular oxygen seen in photosynthesis. Proof of the purpose of magnesium consists of an experiment using a magnesium compound in water and carbon dioxide and sunlight, which forms molecular oxygen without chlorophyll. Proof of the necessity of ultra violet is shown by the appearance of two red fluorescent spectral lines which are formed *only* from electrons returning from the far ultra violet to the first quantum state. The molecular combination, CH_2O is an unstable highly energized compound in the light, losing its energy in darkness to form the carbohydrates at the ground energy level. The entire process of photosynthesis depends upon electrons triggered by light and darkness in a complete cycle.

The mechanism by which green plants produce photosynthesis when exposed to sunlight has evaded solution for over one hundred years. One of the most discussed problems in this connection, is where the energy comes from to split off molecular oxygen from water and carbon dioxide in the presence of sunlight, as the first step in this process. And furthermore, how do green plants synthesize organic compounds from water and carbon dioxide, with the help of sunlight—which is no doubt the most fundamental of all biochemical reactions.

In this connection Rabinowitch (1945) states "what is needed for the success of the photochemical reaction is the capacity of the primarily light-activated molecule to utilize its energy for direct reduction of a compound with a potential negative enough for all subsequent steps to the reaction sequence to go "down hill". He states further in more concrete terms, "if there is a primary photochemical oxidant, its reduced form must be capable of carrying on the reduction of whatever has to be ultimately reduced".

The answer to this problem of the primary photochemical oxidant is the atom of magnesium in chlorophyll whose two valence electrons are removed by sunlight leaving a high positive charge on the magnesium atom as a first step. The second step is the attraction of two oxygen valence electrons to this positive charge, leaving two oxygen atoms with a seventh unpaired electron each. The third step is the union of the two oxygen atoms having an unpaired electron each, to form the oxygen molecule, and the union of the remaining atoms of water and carbon dioxide to form the unstable molecule, namely CH_2O, as the first reduction. Further reduction is also accomplished by the electrons which lose their energy *in the dark* and are therefore the force carrying on the reduction of "whatever has to be ultimately reduced", as the electrons lose their energy at the ground state. In short, it is the *electron* which is the primary unit, performing the whole process through absorption and release of energy.

The following pages give more details of this process and explanations of certain specified related subjects.

Many experiments have been done to find such an oxidant where this energy can be located, —whether in the ATP combination or in other group formations; whether in combinations of visible light with such factors as phosphorylation, the quinone structure, radioactive carbon, and other biochemical combinations.

In all these experiments attempts have been made to solve the problem by considering chlorophyll as a *whole molecule* according to the laws of chemistry at the ground energy level, rather than as separate atoms and electrons energized at specific absorption bands and at change in energy relations in sunlight. That is why the primary photochemical reaction was thought to be due to the splitting of the molecule of water into H atoms and OH radicals, and the dissociation of the molecule into atoms which in turn enter into other combinations.

THE MAGNESIUM ATOM IN CHLOROPHYLL

Many references refer to the magnesium atom of chlorophyll as being connected in some way with photosynthesis. Some of these have endeavored to demonstrate that ultra violet light could be the energy for this primary activation but rejected its importance with the conclusion that it destroys chlorophyll. These include the results of Meier (1936), Rabinowitch (1954), Ruben, Kamen, and Hassid (1940), Arnold (1933), Richter (1934) and others. They used the wave length of 2536.6 Angstroms, produced by the mercury arc or lamp where the intensity of absorption is 10 (the highest in a scale of 1 to 10) as shown by Kayser (1926). However since the solar wave lengths reaching the earth's surface are cut off by the atmosphere at 2900 Angstroms as found by Luckiesh (1922,) Sanderson and Hulbert (1955) and others, the results of experiments using wave lengths at 2536.6 Angstroms would have no application to the process of photosynthesis occurring in nature. Levitt (1954) has also called attention to the magnesium in chlorophyll and states the following, "The energy level for the first ionization potential of magnesium is 7.61 corresponding to a wave length of 162 mμ which is far in the ultra violet and hence not available ordinarily for photosynthesis". This wave length quoted for the first ionization potential is not correct. He may have had access to some figures *in vacuo* sources which are sometimes used for shorter wave lengths instead of those in air. On this basis he rejects the importance of magnesium and reverts back to interpretations through oxidation-reduction potentials.

The magnesium atom has been described as being *un-ionized* according to Rabinowitch (1945). This definition holds for the *stable neutral state at the ground energy level*, where it is loosely held in the center of the chlorophyll molecule, with its two valence electrons as links vibrating in constant motion between certain nitrogen bonds of the four pyrrole rings. This is the description of the un-ionized atom. However when ultra violet light is absorbed by chlorophyll of green plants, the atoms and molecules are no longer in a stable state as found in chemical experiments, but become energized to higher quantum levels and react according to physical laws governing atomic energy, in which *electrons* instead of cations and anions have specific relations. The explanation here presented for photosynthesis shows these results, whereby atoms and their electrons absorb energy at high levels to initiate the primary reaction for excitation by *light*. And conversely, whereby the electrons lose this energy in *the dark* as the various step-wise reductions take place, ending at the stable ground state. The entire process is triggered by electrons of *specific atoms* highly energized at specific wave lengths in the light, and which lose this energy automatically as they drop back to the ground state in the dark. For this reason it is not possible to use laws dealing with stable states of atoms to explain how unstable states react. Details based on changes in energy level of electrons are discussed in the following pages.

The magnesium atom of chlorophyll has absorption bands in the ultra violet region of the solar spectrum at wave lengths from 2915 to 2971 Angstroms which absorb the adequate energy required as shown in Table 1. These absorption bands are within the borderline of 2900 Angstroms where ultra violet light is cut off from the earth's surface. The energy absorbed by the magnesium at this region of the solar spectrum is greater than that holding the two

Table I. Solar wave lengths in the ultra violet spectrum of magnesium reaching the earth's surface*

Angstroms	
2915.5	2965.1
2928.7	2967.8
2936.5	2969.0
2936.8	2971.7
2937.0	

After Moore (1950), Kayser (1926), and Stanley (1911).

valence electrons in the bond structure of the pyrrole ring, and greater than the energy keeping them within the radius of attraction to the nucleus of the magnesium atom. They are therefore removed to an "infinite distance" where the quantum number N= ∞ as described by Hildebrand and Powell (1954), leaving a high positive charge on the magnesium. The energy for removal of these electrons is given at an Ionization Potential of 7.61 electron volts for the first valence electron and 14.97 electron volts for the second as shown in Table II. This high positive charge attracts other electrons. These are from the oxygen atoms of water and carbon dioxide as described in the next section.

The interpretation of "ionization potential" requires an explanation in a later section, since its term in connection with atomic energy has been confused with the word "ionization" as used at the ground energy level in chemistry.

THE OXYGEN ATOM

The oxygen atom in the neutral state has no absorption bands around 2900 to 3000 Angstroms where the transmission of ultra violet wave lengths to the earth's surface ends. An atom will not absorb all energy equally but only a limited number of spectral lines which are specific for that atom. Only such rays as are absorbed by the system can have a photochemical effect, according to Grotthus Law. For this reason it is not possible to separate an electron from the oxygen atom by subjecting water and carbon dioxide to sunlight as attempted in certain previous experiments; nor can oxygen evolution occur without chlorophyll as stated by Rabinowitch (l.c.) The writer has found however, if magnesium nitrate or chloride, for example, is placed in distilled water and carbon dioxide in sunlight, bubbles of oxygen are given off. None such occur in the controls without magnesium. (Unpublished data by Brooks (1965). This demonstrates definitely the relation of oxygen to magnesium.

The neutral atom of oxygen has many absorption bands in the solar spectrum beginning at 3823 Angstroms, according to Kayser (l. c.). This raises the energy of the oxygen and its electrons to a higher *Excitation potential*, but not high enough to the *Ionization potential* which removes the first valence electron. However the high positive charge on the magnesium atom attracts the first valence electrons of the oxygen atom of water, and of the first oxygen atom of carbon dioxide. Since the ionization potential of this oxygen electron is 13.56 electron volts, it loses its energy before reaching 14.97 electron volts, the energy at the magnesium atom. This leaves the charge on magnesium vacant to attract other oxygen electrons, resulting in a constant stream of such electrons to this source, where their energy becomes zero.

The electrons of the second oxygen atoms of carbon dioxide are too firmly held to the carbon bond. Also the second valence electron of oxygen in each case, requires an ionization potential of 35 electron volts for their removal. This energy is not attained by sunlight in this case as shown in Table II.

The result of the removal of one electron from each of these two oxygen atoms, leaves them with a seventh unpaired electron each, instead of the normal eight or octet. In such cases these two atoms, i.e., the one from water and the other from carbon dioxide unite through a co-valent bond to form the octet of molecular oxygen (Hildebrand and Powell, 1954). This is the known structure of all

Table II.   Atomic energy levels of the ultra-violet solar spectrum of
magnesium and oxygen

Atom	Wave-Number-Limit cm^{-1}	Ionization Potential (IP) in Electron volts (ev)
Magnesium I*	61669	7.61
Magnesium II*	121267	14.97
Oxygen I*	109836	13.56
Oxygen II*	283550	35.14

* Roman numerals designate removal of the first or second valence electron.   IP designates energy in electron volts to remove one of the valence
electrons from its atom.   After Moore (1950).

magnetic molecules including molecular oxygen according to Pauling
(1962) and the source of the stream of the oxygen molecules as seen
in green plants in sunlight.

THE FIRST REDUCTION PRODUCT, CH$_2$O.

After the removal of the two oxygen atoms, one each from water
and carbon dioxide; the remaining two hydrogen atoms are attracted
to the carbon atom of CO to fill the vacancy left there when the
first oxygen atom is removed.   This molecule is the well-known,
CH$_2$O, since carbon forms non-polar bonds in which it shares four
pairs of electrons with surrounding atoms, by adding or losing electrons as the case may be.   CH$_2$O is the chemical formula for formaldehyde, and was once considered to be one of the steps in the reduction process of photosynthesis.   However formaldehyde is a *stable*
chemical at the neutral ground state of maximum stability at the
equilibrium attained in chemical systems.   In an energized state as
that produced by sunlight, its atoms and electrons are all *unstable*
and only temporarily united, and lose their energy in the dark
almost instantaneously, combining with other biochemicals to form
the various compounds as they drop back to the stable energy level
in the negative region of the redox potential.   In this way they

form the carbohydrates and sugars which have been described by Calvin and associates (1962). One can therefore conclude that the "force which directs the reduction of a compound to the "down hill gradient" is simply the opposite of high energy absorption, namely high energy loss in the dark, accomplished by *electrons*. This is an example of the electrical property of matter as discussed by Brooks (1963).

VISIBLE LIGHT

The region of the solar spectrum studied by others is between 400 and 900 mμ. Spectral lines of chlorophyll as a molecule and those of the surrounding medium were noted. The importance of visible light in photosynthesis is to energize other atoms beside the magnesium, in order to assist in the absorption of energy at an excitation potential sufficient to use this energy in the formation of the various chemical compounds produced "on the way down" to the negative region of the redox potential at the ground energy level.

Carotenoids which are accessory pigments have no magnesium in their formula and therefore their electrons do not attain the energy level of excitation potentials reached by magnesium. They do assist in the formation of biochemicals as they lose their energy in a dark, but cannot form molecular oxygen nor fluorescence.

THE BACK REACTION

The last step or "back reaction" in photosynthesis is the respiratory mechanism. This is the well-known method of activation of molecular oxygen of the air to combine with hydrogen and carbon activated through the successive enzyme reactions as indicated in the redox potential scale, ending in the formation of water and carbon dioxide to complete the cycle in aerobic plant life. This step takes place at the ground energy level. Warburg et al. (1951)

has previously discussed this as the process of respiration and not as a reversal of photosynthesis, in the case of green plants.

INFRA-RED LIGHT IN PHOTOSYNTHESIS

Photosynthesis in green plants and in the lower forms has certain similarities, inasmuch as their atoms and electrons absorb energy in the light and lose it in the dark. They differ from each other in the energy level at which these electrons become activated. In the dark all atoms and electrons are in the normal state of minimum potential energy. In the light they all absorb energy which depends upon the wave length of the specific atoms present.

When infra-red light alone is used in such low forms the entire reactions take place at the ground energy level, differing in this respect from those of green plants. No ionization potentials are produced in this light, but merely a transfer of electrons to atoms of a slightly more positive charge, which produces the biochemicals at this low energy level, in the negative region of the redox potential.

Considerable research has been done on such forms as purple bacteria, *Spirrilum rubrum*, which grow in a medium of pure hydrogen sulfide or in other organic media, which can be used as hydrogen donors, as glucose and sugars. These forms have bacterio-chlorophyll, which is apparently a reduced form of chlorophyll "a", having two extra hydrogen atoms attached to the pyrrole ring. This produces a shift of the absorption band towards the infra-red region, giving a maximum around 800 to 900 mμ, depending upon the nature of the solution used for measurement, as stated in the volumes of Rabinowitch (l. c.). In such cases sufficient energy is absorbed by the hydrogen atoms and their electrons to be activated to slightly higher levels for the survival of these low forms of organisms.

Table III shows several spectral lines in the far red region of

Table III.   Infra-red wave lengths of magnesium and hydrogen
in the normal state*

Magnesium	Angstroms	Hydrogen
23991		8029 to
23977		6090
23963		covering 34
11054		special lines
9224		

*After Kayser (1926)

the neutral magnesium atom at 9224 Angstroms and beyond this, and also numerous absorption bands of the hydrogen atom from 6962 to 8029 Angstroms.   At this low energy level such enzymes as the SH groups of cysteine-cystine-glutathione systems are active.   This is also the region where glycolysis in lower animals occurs due to incomplete oxidations.

In sulphur bacteria, the energy rises from—.4 volts, that of hydrogen at the ground state, to—.22 volts where cysteine-cystine group of enzymes is active.   It has been found that free sulphur is released as a fine powder by these organisms in their metabolism, as the end product, indicating the limit of oxidations in this case. The light-activated hydrogen forms combinations with other atoms and the sulphur atom is released.   Since sulphur has the tendency to share its electrons to complete the octet, the free element as a fine powder is deposited.   Other examples of organisms growing in this negative redox potential region are discussed by Rabinowitch (l.c.).   *Spirrilum rubrum* is used here as an example.

In conclusion it may be stated that the energy of infrared light is sufficient to maintain the growth of certain lower forms of life, taking place in the negative region of redoxpotential.   It is an example of incomplete oxidation where no oxygen occurs but the whole process consists of electron transfer, where hydrogen is activated rather than oxygen as in green plants.   In these cases no carbon dioxide is produced since the absence of oxygen eliminates this formation.

The term "oxidation" is a misnomer from early days as demonstrated by Clark (1923). When an atom transfers an electron in chemical terms, the atom becomes oxidized, and when it gains an electron, it becomes reduced. The former can take place with or without the participation of oxygen, and the latter with or without the participation of hydrogen, the entire process in either case, depending upon electron transfer.

MONOCHROMATIC LIGHT USED FOR PHOTOSYNTHESIS

Many experiments using monochromatic visible light at different wave lengths have attempted to show that photosynthesis takes place at all regions of the spectrum, including, blue, yellow, green and red light according to Hoover (1937). Zscheile et al. (1943) found fluorescence in the red region of the chlorophyll spectrum when mercury lines of 365, 404, 435, or 546 mμ or white light filtered through red, orange, blue, green or violet filters were used and concluded that photosynthesis takes place at all wave lengths.

In experiments of others broad spectral bands about 200 mμ apart were used with no reference to absorption by corresponding wave lengths of separate atoms. In some experiments according to Rabinowitch (1951) such wave lengths as 365 and 470 mμ in the visible region showed no difference in photosynthesis. The sources for such energy absorption was the mercury arc, Mazda light, the carbon or helium arc and concentrated sunlight. Such wave lengths as stated above have no counterpart in the magnesium atom and therefore are not absorbed by it. They follow Grotthus Law which states that only such rays as are absorbed by the system can possibly have a photochemical effect. However there are certain absorption bands in the mercury arc which are similar to the specific wave lengths in the ultra violet absorption spectrum of magnesium, and which are therefore capable of producing effects. Among these are 294, 292, 291 mμ in both atoms as found in Kayser (l. c.) and Moore (l. c.). Similarities such as these also occur in the helium and carbon arc. These are the bands, among others, in solar radiation reaching the earth's surface, which are effective in photosynthesis.

In all such experiments, the two red fluorescence spectral lines appeared and could not be explained. Inasmuch as this red fluorescence comes only from the return of electrons from the ultra violet region of the spectrum, it is definite proof that these experiments described above with monochromatic light did not eliminate ultra violet light, and that the use of monochromatic filters were without effect. There seems to be no attempt made in these experiments to relate absorption to specific wave lengths of atoms in chlorophyll or in the solvents used.

THE TWO RED FLUORESCENT SPECTRAL LINES IN CHLOROPHYLL

All radiated atoms and electrons are in a continual state of change in energy level as they absorb and release it. When an electron moves from a high quantum state to a lower one, the difference in energy is emitted in the form of light as a new spectral line appears. Transitions from outer energy levels to the second quantum state give rise to the Balmer Series which is the visible light. Such a transition represents electrons returning from higher orbitols, such as those indicated by the third or fourth quantum state. However, the fluorescence occuring at the red spectral region in the first quantum level results *only* from electrons returning from the far out ultra violet region, as stated in Hildebrand and Powell (1954). This is the origin of the two red fluorescent bands in the chlorophyll. It is definite proof of the absorption of ultra violet. It represents the energy of the emitted light at the first quantum level by the two valence electrons of magnesium, which are released at an ionization potential of 7.61 and 14.97 electron volts respectively, and which return to the two different energy levels in the red spectrum at the first quantum level. Since it requires more energy to release the second valence, this one would be able to drop to a lower level which is the infra red region.

Finally, the origin of the two red fluorescent lines indicates that there are two separate electronic transitions rather than two spectral lines in the same band.

There is no relation between the red fluorescent bands of chlorophyll and the red absorption bands of chlorophyll which are mainly those of hydrogen as shown in Table III.

WINDOW GLASS AND TRANSMISSION OF ULTRA VIOLET LIGHT

The question has been raised concerning the region of the solar spectrum which is cut off by window glass, as for example, in green houses. Luckiesh (1922) states that wave lengths at 295 mμ is the limit of transmission by ordinary window glass. Since plants are able to grow in green houses, it seems important to comment on this factor, with reference to absorption bands of the magnesium atom of chlorophyll. Table I shows a number of wave lengths beyond the 295 mμ region which are within the limit specified for transmission through window glass, where magnesium has absorption bands in the ultra violet region.

IONIZATION POTENTIALS

There is a difference in the interpretation of the term "Ionizing and Ionization" potentials when referring to chemical reactions taking place at the ground state and when referring to atoms and electrons which have been "ionized" to higher energy levels, as used in physics. In the former case the atoms and their ions are in stable equilibrium, and their reactions are between cations and anions. In the latter case the entire process is based on *electrons* at higher energy levels. In the latter case the positively charged unit remaining is called an "atom" in accordance to atomic energy terminology, rather than "ionic energy", referring to the element

The term "ionization" is a misnomer as stated by Kayser (l.c.). It was first used by Lockyard who did not understand the difference in those early days, and unfortunately the term still persists. Kayser refers to this change in energy levels of such atoms as though they had become different elements, as successive electrons are lost at successive higher radiation energy. However the loss of each electron from its atom in terms of such radiation physics is still referred to as "ionization potential". Moore (1950) refers to "the atom excited by radiation as absorbing certain wave lengths corresponding to transitions of their outer electrons from lower energy levels to high ones", or to the number of electrons of the "atom" with reference to the "ground state". In the present article on photosynthesis, the term "atom" is used in discussing the reactions of magnesium and oxygen, which is the terminology found in the physical tables of atomic energy levels.

RESUME

The entire photosynthetic cyle is based on atomic energy levels of electrons of specific atoms through changes in sunlight and darkness, rather than through chemical changes in molecules.

The magnesium atom of chlorophyll is un-ionized only at the ground energy level but becomes ionized by absorption of energy by its spectral lines from 2915 to 2971 Angstroms.

A high positive charge is produced on the magnesium atom when its two valence electrons are removed at ionization potentials of 7.6 and 14.97 electron volts respectively.

The magnesium atom is the key to photosynthesis through absorption of ultra violet light at the exact wave length within the boundary of solar light reaching the earth's surface. The limit of solar light reaching the earth's surface is 2900 Angstroms.

The high positive charge produced on the magnesium atom when its two valence electrons are removed attracts the first valence electron of two oxygen atoms, namely, of the oxygen atom of water, and of the first oxygen atom of carbon dioxide.

The ionization potential of the oxygen electrons is less than the charge on the magnesium atom, resulting in the loss of energy by the oxygen electron before it reaches the charged magnesium, where the quantum number, $N = \infty$ and the energy of the electron zero. This leaves the magnesium atom open to attract further oxygen electrons from the same source, resulting in a constant stream of oxygen electrons.

The removal of one electron each from the two oxygen atoms described, leaves a seventh unpaired electron in each case. These unite to form the oxygen molecule seen in photosynthesis.

The remaining atoms unite to form CH_2O which is an unstable molecule at a high energy level, differing in this respect from formaldehyde, a chemical which is stable at the ground state.

In the dark all electrons lose their energy acquired as they revert to the ground energy level and recombine with other atoms and finally form carbohydrates and sugars.

Experiments using monochromatic filters in the visible light are of no avail as they have not eliminated ultra violet light as proven by the appearance of red fluorescence.

The "back reaction" in green plants follows the same steps as are taken in the well-known respiratory mechanism through the activation of enzymes as the light begins to shine again.

The red fluorescence of two spectral lines at the first quantum level can only be produced by electrons falling back from the far out ultra violet region. It is absolute proof that ultra violet light is absorbed when they occur. They have no connection with absorption bands in the red or infra red region.

In low plant forms metabolism takes place at the first quantum level entirely by oxidation-reduction reactions within the negative region of the redox scale. Energy is obtained chiefly by the many hydrogen absorption bands as the hydrogen becomes activated in the light through enzymes, and loses it in the dark. The energy is too low to activate oxygen or to produce carbon dioxide.

Further additional proof of the key role of magnesium in photosynthesis consists of an experiment in which a magnesium compound such as magnesium nitrate, distilled water and carbon dioxide are placed in sulight without chlorophyll. This forms bubbles of oxygen. This agrees also with the results of others showing that the carotenoids which have no magnesium in their molecule cannot produce oxygen and have no red fluorescence.

Explanations are given of the term " ionization" as used in atomic energy as compared with its meaning in chemistry; and of oxidation -reduction to indicate the role of electrons: and concerning transmission of light through window glass.

CONCLUSION

Photosynthesis depends upon the absorption of energy by two specific wave lengths of the magnesium atom in chlorophyll and the ability that ultra violet can be absorbed at the exact energy level.

Molecular oxygen is given off by chlorophyll in sunlight. It can also be given off by a compound of magnesium in distilled water and carbon dioxide when placed in sunlight when no chlorophyll is added.

The proof of absorption of ultra violet light is found when two red fluorescence spectral lines occur in the spectrum of chlorophyll experiments. Such red fluorescence can only occur when electrons which have absorbed energy in the far ultra violet light return to the first quantum state and give off a red spectral line. This proves that ultra violet light is fundamental to photosynthesis and the magnesium is the absorption area for the ultra violet light in the process of photosynthesis.

Photosynthesis in low plant forms takes place in the negative region of the oxidation-reduction scale through changes in energy levels of electrons in light and darkness.

In both cases the electron is the ultimate principle or driving force due to changes in light and darkness.

The entire cycle can be regarded as the basic principle of the life processes.

CONVERSION FACTORS

Table IV gives the conversion factors to show the relation between electron volt quanta which indicates the energy absorbed at the wave-number-limit of the specific atom, and the energy in ergs required for attaining this level.

Table IV. Conversion factors to show energy required to remove the first oxygen electron and the second magnesium electron

	Oxygen	Magnesium
Wave-number-Limit in cm^{-1}	109836	121267
Ionization Potential (IP) in Electron Volts (ev)	13.6	14.97
Energy in Ergs (IP$\times 1.60206 \times 10^{-12}$)	21.7×10^{-12}	23.9×10^{-12}

8806.03 cm^{-1} =	Wave number of one electron volt.
1.60206×10^{-12} ergs =	Energy of one electron volt.

After Pauling (1962)

The wave number of one electron volt is 8806.03 cm^{-1}. By dividing the wave-number-limit of an atom by the wave number of one electron volt, one obtains the number of electron volts required for the ionization potential at various energy levels. In the case of magnesium it is 14.97 and for oxygen it is 13.6 electron volt quanta.

To convert electron volt quanta into ergs, one multiplies the number of electron volt quanta needed for the ionization potential by 1.60206×10^{-12} (which is the energy in ergs for one quantum). This gives 21.7×10^{-12} for oxygen and 23×10^{-12} for magnesium.

All factors for conversion are taken from Pauling (1962).

LITERATURE CITED

Arnold W. (1933). Effects of ultra violet light on photosynthesis. Jour. Gen. Physiology 17:135-144.

Brooks, M. M. (1963). The electrical property of matter, the trigger mechanism controlling cell growth. Protoplasma 57:144-157.

— (1963). Energy release by electron transfer, the trigger mechanism in living cells. The Physiologist 6:147.

— (1965). Magnesium as the key to photosynthesis. Unpublished experiments.

Calvin, M. (1962). The path of carbon in photosynthesis. Science 135:879-889.

Clark, W. M. (1928). Studies in oxidation-reduction. United States Public Health Service, Bulletin 151. Washington, D. C.

Hildebrand, J., and R. E. Powell (1954). Principles of Chemistry. The Macmillan Co., New York.

Kayser, H. (1926). Tabelle der Hauptlinien der Linienspektra aller Elemente. Julius Springer, Berlin.

Levitt, L. S. The role of magnesium in photosynthesis. Science 120:33-35.

Meier, F. E. (1936). Lethal effects of short wave lengths of the ultra violet on the alga *Chlorella vulgaris*. Smithsonian Miscellaneous Collections 95:1-19.

Moore, C. (1950). Atomic Energy Levels. United States Bureau of Standards Pulications, Washington, D. C.

Pauling, Linus (1962). The nature of the chemical bond. Cornell University Press, Ithaca, N. Y.

Rabinowitch, E. I. (1945, 1951, 1956). Photosynthesis. Vols. I and II. Interscience Publishers, New York.

Richter, O. (1934). Neue Beitrage zur Photosynlheses. Denkschr. d. Akademie der Wiss. Math-Nature. Klasse 103:163-209.

Ruben, S., M. D. Kamen, and W. Z. Hassid (1940). Photosynthesis with radioactive carbon. Jour. Amer. Chem. Soc. 62:3443-3450.

Sanderson, J. A., and E. C. Hulbert (1955). In Radiation Biology, edited by A. Hollaender 95-118. McGraw-Hill Co., New York.

Stanley, F. (1911). Lines in the arc spectra of elements. Adam Hilger, 75 Camden Road, London, England.

Warburg, O., D. Burk, and A. L. Schade (1951). Extensions of photosynthetic explanations. Symp. Soc. Exp. Biol. 5:306.

Zscheille, F. P., and D. G. Harris (1943). Studies on the fluorescence of chlorophyll. Jour. Physical Chem. 47:623-637.

ADDENDA

A few additional experiments are herewith given to show more clearly the differences between the energy of chemical reactions and the energy of an electromagnetic force such as that controlling photosynthesis.

In the experiments of Hill and Scarisbrick (See Rabinowitch, 1945) molecular oxygen was produced when potassium ferric oxalate in water was added to solutions of chlorophyll in the place of carbon dioxide. Proof of oxygen evolution was demonstrated when hemoglobin, added to the experiment was changed from the reduced form to bright red oxyhemoglobin. In this case the oxalate radical (C_2O_4) was split into carbon dioxide radicals by solar energy, and oxygen was produced by photosynthesis as described in this paper.

Hill and Scarisbrick (l. c.) also found that yeast cells added to chlorophyll solutions, without the addition of carbon dioxide, was effective in producing molecular oxygen. This was evidently due to the fermentation of yeast which gives off carbon dioxide.

In several unpublished experiments of the writer it was demonstrated that aqueous solutions of potassium oxalate without the addition of chlorophyll or carbon dioxide, produced bubbles of carbon dioxide in sunlight. None are formed in the dark.

When an aqueous solution of potassium ferric oxalate is placed in sunlight without chlorophyll, the mixture gradually assumes a yellow colloidal consistency which forms a precipitate and after further exposure produces a reddish brown precipitate, which is insoluble and designated as Fe_2O_3, which is soluble in hydrochloric acid. This reaction does not take place in the dark and is not reversible, and the potassium ferric oxalate remains as a clear liquid. These experiments show that energized atoms form temporary bonds with atoms as they lose their energy in the dark.

Previous experiments on photosynthesis have attempted to solve the problem of energy activation by chemical methods where the atom and electron are at the normal state. This paper shows that the electromagnetic force of light produces energized atoms and electrons which are different than those found in normal chemical reactions. Certain facts in this connection have not been recognized as follows;

1. The magnesium atom and its role in chlorophyll as the key to the origin of the energy.

2. Ultra violet light absorbed at definite wave lengths by the magnesium atom.

3. The limits of ultra violet light reaching the earth's surface have been considered at 3000 Angstroms instead of 2900. Experiments using wave lengths at 2536 Angstroms, which killed the chlorophyll, were used as a basis for eliminating ultra violet light, when this wave length did not even reach the earth's surface.

4. The use of whole molecules instead of the atom and its electron as the unit of experimentation.

5. Energy of high power from 10^4 to 10^{10} ergs were used without the realization that the energy in photosynthesis was in the region of 10^{-10} ergs as given for *electromagnetic* force of atomic energy laws.

Appendix B –. Matilda Moldenhauer Brooks, *Journal of General Physiology* 50 (1967), 2508: "Photosynthesis without Chlorophyll

Photosynthesis without Chlorophyll

MATILDA M. BROOKS, Department of Physiology, University of California, Berkeley, California

The "force" which controls photosynthesis is the diurnal rhythm of day and night, forming molecular oxygen in the day and carbohydrates at night, as previously published. The key is the magnesium of chlorophyll and its specific wavelengths in the ultraviolet region of the spectrum. The proof of this consists of using a magnesium compound dissolved in water containing carbon dioxide, in the absence of chlorophyll or living green plant cells, and placed in sunlight. This combination gives molecular oxygen in sunlight and glucose when transferred to darkness. Many critics state that photosynthesis of green plants also takes place in red light alone. In these experiments the tungsten filament was used instead of daylight. However, the tungsten atom has wavelengths in the ultraviolet region which are similar to those of magnesium and solar light, so that tungsten was simply a substitute for solar light. A further proof of the necessity of ultraviolet light is red and infrared fluorescence, which can *only* come from electrons returning from the far ultraviolet region of the spectrum to the ground energy level. The well-known $(CH_2O)_n$ is a polyatomic free radical, highly energized in the light, and formed by a temporary union of the atoms of water and carbon dioxide remaining after 1 oxygen atom each from water and carbon dioxide has been removed. Details have been published. It is the *origin* of the plant carbohydrates, as its electrons lose their energy in the dark and split off simple free radicals at different energy levels to form the basic plant acids and glucose as "precipitates." The test for glucose using Benedict's solution confirms this. Combinations of these radicals with the minerals of the soil form the "back-reaction" or respiration, an oxidation-reduction system, by which oxygen is activated to form water and carbon dioxide. It is the *source of supply* for photosynthesis, rather than the air and water, as generally believed. This completes the automatic cycle triggered by day and night through atomic energy of electrons.

Appendix C – Matilda Moldenhauer Brooks, *Advancing Frontiers of Plant Science* 23 (May 1969), 55-63, "Further Interpretations of Photosynthesis"

FURTHER INTERPRETATIONS OF PHOTOSYNTHESIS

Matilda M. Brooks

Department of Physiology, University of California

Berkeley, California

SUMMARY

Several problems concerning photosynthesis which have been misunderstood—are explained in detail. These include the transformation of radiant energy (red fluorescence or emission) into atomic energy, as based on the Einstein Frequency Law with reference to photosynthesis; that the carbon dioxide and water used for photosynthesis is not correctly stated and is proven in experiments given here that the carbon dioxide does not come from the air but is produced along with the water used in photosynthesis, as the final step of respiration; the relation of the three chlorophylls to each other with reference to the number of hydrogen atoms of chlorophyll and this effect on energy levels; the formation of the basic plant acids and glucose from the high energy of the polyatomic free radical, CH_2O when it loses its electrons in the dark. Also two experiments showing the production of glucose without chlorophyll, by substituting magnesium for chlorophyll, and the *in vivo* formation of glucose in the marine alga *Valonia*; and the proof that cyanide does not stop photosynthesis, but stops respiration upon which photosynthesis depends for its carbon dioxide and water.

ORIGIN OF THE CARBON DIOXIDE AND WATER USED FOR PHOTOSYNTHESIS

The general belief concerning the origin of the carbon dioxide used by green plants for photosynthesis is that *it comes from the air.* However the writer has demonstrated that carbon dioxide must be added to water in experiments where magnesium was used in place of chlorophyll to produce molecular oxygen in sunlight (Brooks, 1967). Controls in which a magnesium solution was placed in sunlight exposed to air without the addition of carbon dioxide were negative. The following explanation is given of the origin of the carbon dioxide and water for photosynthesis, which will show that the *plant produces its own carbon dioxide and water* in the *chloroplasts* through the process of respiration, and does not get it from the air.

Carbon dioxide is very soluble in water and is always present in living cells whether growing directly in water which is absorbed by aquatic plants, or in moist soils transferred through the roots. This is not the water or carbon dioxide used for photosynthesis, but the basis of the respiratory system, where the final step of the well-known oxidation-reduction system activates the cytochromes. These iron-containing enzymes activate the oxygen to combine with the carbon of the foodstuffs to form carbon dioxide and with the hydrogen from the same source to form water. This is the *source of carbon dioxide* and *water* used in photosynthesis. This takes place in the chloroplasts where the carbon dioxide and water are in direct contact with the magnesium of chlorophyll. It is this closely formed unit which produces molecular oxygen in sunlight as described by Brooks (1966). The molecular oxygen is eliminated through the stomata of the leaves, while the remaining atoms from water and carbon dioxide remain in the chloroplasts to produce the carbohydrates.

EXPERIMENTAL

For further proof of the source of carbon dioxide and water for photosynthesis, an experiment was performed in which the process of photosynthesis was separated from respiration. In this experiment molecular oxygen was produced in the same manner as that of photosynthesis by the plants by using a magnesium compound instead of chlorophyll and placed in sunlight as described by Brooks, l.c. When a chemical which is known to stop respiration was added to this solution, no effect on the production of molecular oxygen occurred. In this case, 2 cc. of a 1 percent of a solution of potassium cyanide was added and bubbles of oxygen continued to appear. The

Chart I.   Photosynthesis with or without chlorophyll

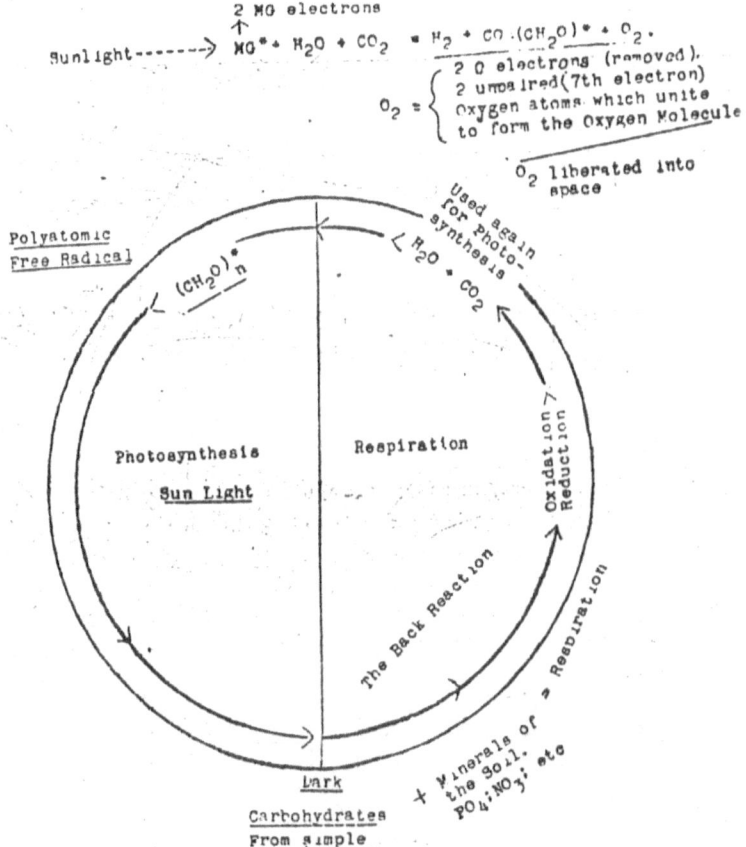

2 MG electrons

Sunlight ------> MG* + H_2O + CO_2 = H_2 + CO .(CH_2O)* + O_2 .

2 O electrons (removed).

O_2 = { 2 unpaired(7th electron)
oxygen atoms which unite
to form the Oxygen Molecule

O_2 liberated into space

Used again for Photo-synthesis

H_2O + CO_2

Polyatomic Free Radical

(CH_2O)*$_n$

Oxidation 7 Reduction

Photosynthesis

Sun Light

Respiration

The Back Reaction

= Respiration

Dark

Minerals of the soil. PO_4; NO_3; etc

Carbohydrates
From simple
Free Radicals;
CH,CHO,OH,(C),
CH_2,COOH

Plant Acids
Glucose

A complete automatic cycle directed by electrons through the electro-magnetic
energy of solar light according to atomic energy laws.

usual test for oxygen as previously described proved the presence of oxygen.

Therefore it is evident that the addition of cyanide did not affect photosynthesis as described in experiments quoted by Rabinowitch (1945) but acted upon respiration and indirectly upon photosynthesis. It is generally known that cyanide inactivates the iron-containing enzymes of the oxidation-reduction system and therefore stops respiration, which produces the carbon dioxide and water as a final step and which is used for photosynthesis.

The conclusion therefore is, since cyanide stops respiration and does not stop photosynthesis when separated from respiration, that the source of the carbon dioxide and water used by the plant comes from respiration and not from the air.

These closely united reactions all within the confines of the chloroplast are diagramed in Chart I to illustrate the completely inter-related cycle which operates as an automatic unit.

ORIGIN OF THE BASIC PLANT ACIDS AND GLUCOSE

The origin of the basic plant acids and glucose has been published in a preliminary abstract by the writer (1967). Further details are given in Table 1. This shows how the polyatomic free radical, CH_2O, forms simple free radicals as its electrons lose their energy in the dark. These are CH, CH_2, C, CHO and $COOH$ which recombine at different energy levels depending upon the energy of the electron which is lost as it reaches the ground state. These stable formations are oxalic, tartaric, malic, and citric acids and glucose, with the latter at the lowest redox potential. This is the formation of the plant carbohydrates, of green plants. Rice (1966) gives an account of the production and reactions of free radicals in outer space.

EXPERIMENTAL PROOF OF THE METHOD OF GLUCOSE PRODUCTION
BY PLANTS

Proof of the formation of glucose without chlorophyll is found by using a solution of a magnesium compound such as the nitrate or chloride dissolved in water containing carbon dioxide. This is placed in sunlight for a few hours and then transferred to darkness. When Benedict's solution is added to this solution and heated, it produces a reddish brown precipitate which is the test for glucose. No such precipitate is formed in any of the controls consisting of magnesium or carbon dioxide separately, or in combinations of one or two only in water or in darkness or all three entirely in darkness.

Table 1. Formation of Free Radicals Produced by $(CH_2O)_n$ as the Electrons Lose their Energy in the Dark.

Free Radicals CH, CH_2, CHO, OH, COOH, in different combinations to produce Basic Acids of Plants and Glucose and Starch.

Acid	Chemical Formula
Oxalic	COOH
	COOH
Tartaric	CH (OH). COOH
	CH (OH). COOH
Malic	CH (OH). COOH
	CH_2 (OH). COOH
Citric	CH_2. COOH
	C (OH). COOH
	CH_2 COOH
Glucose	CH_2 (OH). (CH. OH). (CHO)
Starch	High polymeric form of Glucose

Measured concentrations of chemicals are not required because this is not a chemical experiment dealing with ions at the ground energy level, but deals with electrons which are energized and not subject to such measurements.

Additional evidence of the origin of plant carbohydrates consists of an experiment recently performed at the Bermuda Biological Laboratory using the marine green alga *Valonia*. This is a sac-like plant containing clear sap which can be easily extracted free from debris. In former experiments on *Valonia* it was found that the oxidation-reduction indicators having a positive redox potential penetrated readily and became reduced. No reason for this was discovered. Brooks (1932). However in the light of studies on photosynthesis which included the explanation of the energized free radical, CH_2O, the reduction can now be attributed to the presence of glucose in the sap, and the negative redox potential which it produces. In the present experiment, sap was removed from *Valonia* by puncturing into the interior of the *Valonia* with a fine pointed pipette. When Benedict's solution was added to this sap and heated, a reddish brown precipitate was produced. This is the test for glucose. It is therefore conclusive that glucose is formed directly from CH_2O by solar energy which is lost in the dark. No tests for the basic plant acids were made. Table I however indicates how they were formed as the electrons lost their energy at different levels, with glucose at the lowest negative potential.

APPLICATION OF THE EINSTEIN FREQUENCY LAW

In a previous publication by the writer (1966) the source of solar energy for photosynthesis was taken from tables on *emission spectra* of the magnesium atom of chlorophyll. This statement evidently requires further interpretation in explaining how *absorption* of such energy can be obtained from *emission* wave lengths.

In all atoms there exists a fundamental orbit in which the electron is stable continuously at a minimum energy level. However, in cases where radioactive disintegration occurs, the electron must give up energy. If the atom is isolated and no direct mechanical energy transferance is involved, the emission of an electromagnetic wave length offers the only possibility of liberating energy.

In photosynthesis such *emission* is the red fluorescence. This has its origin in the radioactive disintegration of atoms in the far ultra violet region of *outer space*. As some of these electrons reach the surface of the earth, collision with other atoms occurs, and a wave length is emitted. This is the case when such an electron collides with the magnesium atom which has the specific wave lengths for *absorbing* this *emitted* energy. This illustrates the principle of the *transformation of radiant energy* (fluorescence) into *atomic* energy. It is explained by the Einstein Frequency Law which was later extended by Bohr to include spectral lines. It depends upon the energy differences between two orbits. In the case of photosynthesis, the two orbits are that of the neutral atom of magnesium and the high energy of the emitted wave lengths which was *absorbed* by the specific wave lengths of magnesium between 2900 and 2995 Angstroms (Brooks, l.c.).

The result of this absorption is the complete removal of the two valence electrons of the magnesium, leaving a high positive charge on the magnesium atom to attract oxygen electrons from water and carbon dioxide as stated by Brooks (l.c.) and the origin of molecular oxygen of plants.

This explains how the first primary reduction in photosynthesis occurs. It cannot be explained by biochemistry which deals with atoms and molecules at the ground energy level, but depends upon the high energy sources of outer space as described in astrophysics.

Photosynthesis is in the same dimension of energy coming under the heading of astrophysics. Other examples of this type are the experiments of Lord Rutherford who duplicated the energy at the sun's surface. This enormous force used disrupted the nucleus of the nitrogen atom by bombardment with alpha particles. However in the case of photosynthesis, the infinitesimal energy level of solar light reaching the surface of the earth is only sufficient to affect the outer electrons of the magnesium atom by removing them completely from the nucleus of the magnesium This leaves a high positive charge on the magnesium to

This leaves a high positive

which oxygen electrons are attracted from water and carbon dioxide to form the molecular oxygen given off by the plants as described in detail by Brooks, l.c. In both cases the principle is the same, namely solar radiation. The difference in reaction depends merely upon the level of energy of the region from outer space.

THE THREE CHLOROPHYLLS OF PLANTS

In green plants chlorophyll has two forms which are described as "a" and "b". The former has two more hydrogen atoms than the latter. In low plant forms a third chlorophyll exists, called bacteriochlorophyll, which has two more hydrogen atoms than "a". According to the publications of Clark (1928) the addition of hydrogen atoms signifies a change in the redox potential to a more negative value, so that the lowest energy level of these forms is bacteriochlotophyll. In addition Rabinowitch (1945) states that the latter form can be oxidized by removing two atoms of hydrogen, which is further proof that it is a reduced form of "a". Examples of such differences are found in the products formed in the dark at specific energy levels due to the presence of "a" and "b". For example glucose has the most negative potential of the plant carbohydrates and is formed by "shade" plants, while starch is formed in "sun" plants where no "a" occurs according to the literature, but where "b" occurs. In other cases where intermediate grades exist, percentage of both kinds varies, indicating specific required energy levels. Other examples can be given where certain aquatic plants have only "a". In this article attention is merely drawn to differences in energy levels as the *cause* of the "precipitates" which the plant produces in the dark.

In contrast to green plants where no oxygen nor carbon dioxide are produced, such as the sulphur bacteria growing in red light, the final product is sulphur instead of oxygen and water. This is precipitated out in the leaves as a fine powder at a redox potential which is much lower than that for glucose, due to the low energy available of visible light.

LITERATURE CITED

Brooks, M.M. (1932). The penetration of certain oxidation-reduction indicators into different species of *Valonia*. Protoplasma 17:89-96.

— (1966). Atomic energy levels explaining photosynthesis. Advancing Frontiers of Plant Sciences 16:51-70.

— (1967). Photosynthesis without chlorophyll. Journ. Gen. Physiol. 50:2508. Preliminary abstract.

— (1968). Molecular oxygen and glucose without chlorophyll. Federat. Proc. 27:508. Preliminary abstract.

Clark, W. M. (1928). Studies on oxidation-reduction. U.S. Public Health Service, Washington, D.C.

Rabinowitch, E.I. (1945). Photosynthesis. I. Interscience Publishers, New York.

Rice, F. O. (1966). Production and reactions of free radicals in outer space. Amer. Scientist 45:158-169.

INDEX